Other Titles in This Series

(See the AMS catalog for earlier titles)

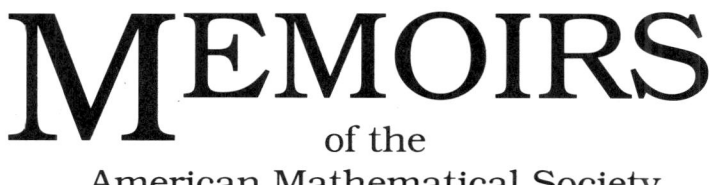

MEMOIRS
of the
American Mathematical Society

Number 580

Reductive Subgroups
of Exceptional
Algebraic Groups

Martin W. Liebeck
Gary M. Seitz

May 1996 • Volume 121 • Number 580 (end of volume) • ISSN 0065-9266

American Mathematical Society
Providence, Rhode Island

1991 *Mathematics Subject Classification.*
Primary 20G15, 20E07.

Library of Congress Cataloging-in-Publication Data

Liebeck, M. W. (Martin W.), 1954–
  Reductive subgroups of exceptional algebraic groups / Martin W. Liebeck, Gary M. Seitz.
    p. cm. – (Memoirs of the American Mathematical Society, ISSN 0065-9266; no. 580)
  "Volume 121, number 580 (end of volume)."
  Includes bibliographical references.
  ISBN 0-8218-0461-8
  1. Linear algebraic groups. 2. Lie algebras. I. Seitz, Gary M., 1943–   . II. Title.
III. Series.
QA3.A57   no. 580
[QA179]
510 s–dc20
[512′.2]
                                                                                              96-4542
                                                                                              CIP

Memoirs of the American Mathematical Society

This journal is devoted entirely to research in pure and applied mathematics.

Subscription information. The 1996 subscription begins with Number 568 and consists of six mailings, each containing one or more numbers. Subscription prices for 1996 are $391 list, $313 institutional member. A late charge of 10% of the subscription price will be imposed on orders received from nonmembers after January 1 of the subscription year. Subscribers outside the United States and India must pay a postage surcharge of $25; subscribers in India must pay a postage surcharge of $43. Expedited delivery to destinations in North America $30; elsewhere $92. Each number may be ordered separately; *please specify number* when ordering an individual number. For prices and titles of recently released numbers, see the New Publications sections of the *Notices of the American Mathematical Society.*

Back number information. For back issues see the *AMS Catalog of Publications.*

Subscriptions and orders should be addressed to the American Mathematical Society, P. O. Box 5904, Boston, MA 02206-5904. *All orders must be accompanied by payment.* Other correspondence should be addressed to Box 6248, Providence, RI 02940-6248.

Copying and reprinting. Individual readers of this publication, and nonprofit libraries acting for them, are permitted to make fair use of the material, such as to copy a chapter for use in teaching or research. Permission is granted to quote brief passages from this publication in reviews, provided the customary acknowledgement of the source is given.

Republication, systematic copying, or multiple reproduction of any material in this publication (including abstracts) is permitted only under license from the American Mathematical Society. Requests for such permission should be addressed to the Assistant to the Publisher, American Mathematical Society, P. O. Box 6248, Providence, RI 02940-6248. Requests can also be made by e-mail to `reprint-permission@ams.org`.

Memoirs of the American Mathematical Society is published bimonthly (each volume consisting usually of more than one number) by the American Mathematical Society at 201 Charles Street, Providence, RI 02904-2213. Second-class postage paid at Providence, Rhode Island. Postmaster: Send address changes to Memoirs, American Mathematical Society, P. O. Box 6248, Providence, RI 02940-6248.

HB 5/96

Contents

Abstract

Let G be a simple algebraic group of exceptional type over an algebraically closed field of characteristic p. We determine the embeddings of arbitrary closed connected semisimple subgroups in G, under some mild restrictions on p: every such subgroup is embedded in an explicit way in a "subsystem subgroup" of G (that is, a semisimple subgroup which is normalized by a maximal torus of G). The proof is based on Theorem 1, which states that if the reductive subgroup X lies in a parabolic subgroup $P = QL$ of G, with unipotent radical Q and Levi subgroup L, then some conjugate of X lies in L. We also use Theorem 1 to prove that $C_G(X)$ is always reductive.

Other results in the paper concern the actions of simple closed connected subgroups X of G on the Lie algebra $L(G)$ of G. Following Dynkin, a labelled diagram is associated with each such subgroup. We prove that when X is of type A_1 the labelled diagram determines X up to conjugacy in Aut G; and we establish a similar, though weaker, result when X has rank 2 or more. Further results give information about centralizers and composition factors of X acting on $L(G)$.

Key words and phrases: algebraic groups, exceptional groups, reductive groups, connected subgroups, Lie algebras

Introduction

This paper is a contribution to the study of the subgroup structure of simple algebraic groups of exceptional type. The maximal closed connected subgroups of these groups were determined in [Se2], subject to some mild restrictions on the characteristic p of the underlying field. Here we take the study further, and investigate arbitrary closed connected reductive subgroups X of an exceptional algebraic group G, again with mild characteristic restrictions (in particular, $p = 0$ or $p > 7$ covers all the restrictions).

We obtain results which determine the embeddings of arbitrary closed connected semisimple subgroups in G. We show that if X is such a subgroup, then X is embedded in an explicit way in a "subsystem subgroup" of G - that is, a semisimple subgroup which is normalized by a maximal torus of G. Subsystem subgroups are constructed naturally from subsystems of the root system of G; this therefore determines the embedding of X in G. As a consequence, when $p = 0$ there are only finitely many conjugacy classes of such subgroups X, whereas there are infinitely many when $p > 0$. The connection with subsystem subgroups is useful in various ways. For example, it is very helpful in finding centralizers of subgroups and in restricting representations.

We present tables which give all the conjugacy classes of simple subgroups X of G of rank at least 2, their connected centralizers, and their actions on $L(G)$, the Lie algebra of G. For subgroups X of type A_1, we associate with each such subgroup a labelled Dynkin diagram, and prove that the conjugacy class of X is determined by its labelled diagram.

Our proofs are based on Theorem 1, which states that if the reductive subgroup X lies in a parabolic subgroup $P = QL$ of G, with unipotent radical Q and Levi subgroup L, then some conjugate of X lies in L. We also use this result to prove that $C_G(X)$ is always reductive.

For simple subgroups X we establish that with essentially one exception, X is determined up to (Aut G)-conjugacy by its composition factors on $L(G)$; and that if X is of rank at least 2, and p is a good prime for G, then $C_{L(G)}(X) = L(C_G(X))$.

Some of our proofs require detailed information concerning the restrictions of certain G-modules to various subgroups of G, such as maximal rank subgroups. Many results of this type can be found in Section 2.

We now state our results in detail. Throughout, let G be a simple algebraic group of exceptional type over an algebraically closed field K of characteristic p. In order to specify our assumptions on p, we define, for certain simple subgroups X of G, an integer $N(X, G)$, as given in the following table.

Received by the editors April 13, 1994
Second author supported by an NSF grant and an SERC Visiting Fellowship

	$G = E_8$	E_7	E_6	F_4	G_2
$X = A_1$	7	7	5	3	3
A_2	5	5	3	3	
B_2	5	3	3	2	
G_2	7	7	3	2	
A_3	2	2	2		
B_3	2	2	2	2	
C_3	3	2	2	2	
B_4, C_4, D_4	2	2	2		

For example, $N(A_2, E_7) = 5$, and so on. (This is the table of [Se2, Theorem 1], with a few additional entries.) If (X, G) is not in the table, set $N(X, G) = 1$. And if X is a non-abelian closed connected reductive subgroup of G, and $X' = X_1 ... X_t$, a commuting product of simple groups X_i, then we define

$$N(X, G) = \max(N(X_i, G) : 1 \le i \le t)$$

In particular, if $p > 7$ then $p > N(X, G)$ for all X, G.

Theorem 1 *Let X be a closed connected reductive subgroup of G, and assume that either $p = 0$ or $p > N(X, G)$. Suppose that X lies in a parabolic subgroup $P = QL$ of G, with unipotent radical Q and Levi subgroup L. Then all closed complements to Q in the semidirect product QX are Q-conjugate. In particular, X is Q-conjugate to a subgroup of L.*

In Section 3 we also obtain a version of Theorem 1 when G is a classical group of small rank (see Theorem 3.8).

The proof of Theorem 1 is based on the fact that Q has a filtration by particular high weight modules for L. Choosing P to be minimal (subject to containing X), we find the embedding of X in L modulo Q, so we can restrict each of these modules to X. Then if V is a composition factor of such a restriction, we show that with a few exceptions, all closed complements to V in the semidirect product VX are V-conjugate, and the desired conclusion follows from this. The machinery for carrying out this proof is developed in Sections 1 and 2. The proof is similar in spirit to those of [LS2, Theorems 2.1 and 6.1], but the representation theory involved is much more complicated, and a great deal of calculation is required.

The following result is an immediate consequence of Theorem 1 (compare [LS2, Theorem 8.1].

Corollary *Let X be a closed connected reductive subgroup of G, and assume that $p = 0$ or $p > N(X, G)$. Suppose that X normalizes a closed unipotent subgroup U of G. Then all closed complements to U in the semidirect product UX are G-conjugate.*

To deduce this from Theorem 1, observe that by [BT, 3.12], UX lies in a parabolic subgroup QL of G with $U \leq Q$; then any two closed complements to U in UX are also closed complements to Q in QX, so Theorem 1 ensures that these complements are Q-conjugate.

Taken together with [Se2, Theorem 1], Theorem 1 leads to a description of all closed semisimple subgroups of G (see Theorems 5 and 7 below). The next result is also a consequence of Theorem 1.

Theorem 2 *Let X be a closed connected reductive subgroup of G, and assume that $p = 0$ or $p > N(X, G)$. Then*

(i) $C_G(X)^0$ is reductive;

(ii) if $p > 0$ then $O_p(C_G(X)) = 1$ (where $O_p(C_G(X))$ denotes the largest normal p-subgroup of $C_G(X)$);

(iii) if X is semisimple, then the rank of $C_G(X)$ is equal to the maximal co-rank among subsystem subgroups of G containing X.

We also establish a result on centralizers of non-connected reductive subgroups (see Corollary 4.5).

The determination of all closed simple subgroups of G leads to the next result.

Theorem 3 *Let X be a simple closed connected subgroup of G with $\mathrm{rank}(X) \geq 2$, and assume that either $p = 0$ or p is a good prime for G and $p > N(X, G)$. Then*

$$C_{L(G)}(X) = L(C_G(X)).$$

Remarks 1. The conclusion of Theorem 3 has been shown to hold also for $X = A_1$ in [LT].

2. The analogues of Theorems 2 and 3 for classical groups are not in general true. For example, if $G = SL(V)$ and X is a subgroup of G such that V is indecomposable for X with composition series $0 < V_1 < V_2 < V$, where $V_1 \cong V/V_2$, then $C_G(X)$ is not reductive. And if $G = SL_n$ with $p = n$, then $C_{L(G)}(G) \neq L(C_G(G))$.

The next theorem and its corollary concern the connection between (Aut G)-conjugacy and linear equivalence on $L(G)$ for subgroups of G.

Theorem 4 *Let X_1 and X_2 be closed connected simple subgroups of G of the same type, and assume that $p = 0$ or $p > N(X_1, G)$. Suppose that X_1 and X_2 have the same composition factors on $L(G)$ (counting multiplicities). Then either X_1 is conjugate to X_2 in Aut G, or $G = E_8$ and $X_1 \cong X_2 \cong A_2$, with both X_1 and X_2 lying in subsystem groups $D_4 D_4$ and projecting irreducibly in each factor.*

Corollary *Let X_1, X_2 be closed connected simple subgroups of G, and assume that $p = 0$ or $p > N(X_1, G)$. If X_1 and X_2 are conjugate in $GL(L(G))$, then either they are also conjugate in $\operatorname{Aut} G$, or $G = E_8$ and $X_1 \cong X_2 \cong A_2$ lying in subsystem groups $D_4 D_4$.*

Remarks 1. The cases with $G = E_8$ and $X_1 \cong X_2 \cong A_2$ lying in $D_4 D_4$ really are exceptions to the conclusion of Theorem 4 and the corollary: E_8 has two conjugacy classes of such subgroups A_2 which have the same composition factors on $L(E_8)$.

2. Again, a result like Theorem 4 is not in general true for subgroups of classical groups. For example, let $p > 0, q = p^a > 1$ and let $X_1 \cong X_2 \cong SL(W)$, where W is a vector space of dimension $m \geq 3$ over the algebraically closed field K of characteristic p. Embed X_1 and X_2 in $D = SL_{m^2}(K)$ via the modules $W \otimes W^{(q)}$ and $W^* \otimes W^{(q)}$ respectively, (where $W^{(q)}$ is a Frobenius q-power twist of W). Then X_1 and X_2 are not $\operatorname{Aut} D$-conjugate, but have the same composition factors on $L(D)$.

We come now to the theorems which describe the embeddings of arbitrary semisimple closed connected subgroups of G. Following [LS2], we define a *subsystem subgroup* of G to be a semisimple subgroup which is normalized by a maximal torus of G. (Except for a few cases where $(G, p) = (F_4, 2)$ or $(G_2, 3)$, subsystem subgroups correspond to closed subsystems of the root system of G; in the exceptions, subsystems of the dual of a closed subsystem are also allowed. The closed subsystems are listed in [Ca].) In order to state the results, we need to make one further definition.

Definition Let $Y = Y_1 \ldots Y_k$ be a commuting product of simple algebraic groups Y_i, and let X be a closed semisimple subgroup of Y. For each i, let $\hat{Y}_i$ be the simply connected cover of Y_i. If A is a subgroup of Y, write $\bar{A}$ for the group $AZ(Y)/Z(Y)$, and for $i = 1, \ldots, k$ let $\pi_i : \bar{X} \to \bar{Y}_i$ be the ith projection map. We call the connected preimage of $\bar{X}\pi_i$ in Y_i the *projection* of X in Y_i. We say that X is *essentially embedded* in Y if the following hold for all i:

(i) if Y_i is of classical type, with natural module V_i (taken to be the natural $2n$-dimensional symplectic module if $(Y_i, p) = (B_n, 2)$), then either the projection of X in Y_i lifts to a subgroup of $\hat{Y}_i$ which is irreducible on V_i, or $Y_i = D_n$ and the projection of X in Y_i lies in a natural subgroup $B_r B_{n-r-1}$ for some $r \geq 0$, irreducible in each factor with inequivalent representations;

(ii) if Y_i is of exceptional type, then the projection of X in Y_i is either Y_i or a maximal connected subgroup of Y_i not containing a maximal torus (and hence is given by [Se2, Theorem 1]).

Note that in (i) of the above definition, if Y_i is of type D_4 then we allow V_i to be any of the three irreducible 8-dimensional modules with high weights $\lambda_1, \lambda_3, \lambda_4$.

Theorem 5 *Let X be a closed connected semisimple subgroup of G, such that every simple factor of X has rank at least 2, and assume that $p = 0$ or $p > N(X, G)$. Choose a subsystem subgroup Y of G, minimal subject to containing X (possibly $Y = G$ of course). Then either*

(i) X is essentially embedded in Y, or

(ii) $Y = G = E_6$, $p = 7$, $X = G_2$ and $X < F_4 < G$ with X maximal in F_4.

The (Aut G)-conjugacy classes and connected centralizers of the simple subgroups X (of rank at least 2) are given in Tables 8.1-8.5 in Section 8; also given are the minimal subsystem subgroups containing X and the composition factors of the restrictions $L(G) \downarrow X$.

Remarks (1) The essential embeddings in (i) can be completely determined using results in Section 2 (indeed, we carry this out in justifying the information given in Tables 8.1 - 8.5).

(2) For the convenience of the reader, there are two further tables, 8.6 and 8.7, in Section 8. For G simply connected of type E_7 or E_6, and X a simple subgroup of rank at least 2, these tables give the restrictions $V_{56} \downarrow X$ and $V_{27} \downarrow X$, where V_{56} and V_{27} are the 56- and 27-dimensional modules $V_{E_7}(\lambda_7)$ and $V_{E_6}(\lambda_1)$, respectively.

The next result concerns subgroups A_1 of G. In Section 6 we define a labelled diagram associated to any subgroup A_1 of G; this is the Dynkin diagram of G, with nodes labelled by various non-negative integers which are determined by the set of weights on $L(G)$ of a maximal torus of the subgroup.

Theorem 6 *Assume that $p = 0$ or $p > N(A_1, G)$. Then any subgroup A_1 of G is determined up to conjugacy in Aut G by its labelled diagram.*

We shall see in Section 6 that subgroups of type A_1 with the same composition factors on $L(G)$ have the same labelled diagram (see Theorem 6.3). Thus Theorem 6 is closely related to Theorem 4 in the A_1 case.

Theorem 6 is proved using the next result, which is the analogue of Theorem 5 for subgroups with a factor A_1.

Theorem 7 *Let X be a closed connected semisimple subgroup of G with a factor A_1, and assume that $p = 0$ or $p > N(A_1, G)$. Then one of the following holds:*

(i) there is a subsystem subgroup Y of G containing X, such that Y is a product of classical groups and X is essentially embedded in Y;

(ii) there is a subgroup $Y_0 = F_4, E_6, E_7$ or E_8 of G, and a semisimple subgroup Y_1 of $C_G(Y_0)$, such that either

(a) $X = Y_0 Y_1$, or

(b) X is essentially embedded in ZY_1, where Z is a maximal connected subgroup of Y_0 not containing a maximal torus;

(iii) $G = G_2$, $X = A_1$ and X is maximal in G.

Remark In (ii) of the theorem, the possibilities for $Y_0 C_G(Y_0)$ are given by Theorem 5, and those for Z by [Se2, Theorem 1]. They are:

Y_0	possibilities for Z	$C_G(Y_0)$ $(G = E_8, E_7, E_6, F_4)$
F_4	$A_1, A_1 G_2, G_2$	$G_2, A_1, 1, 1$(resp.)
E_6	$A_2, G_2, A_2 G_2, C_4, F_4$	$A_2, T_1, 1, -$
E_7	$A_1, A_2, A_1 A_1, A_1 G_2, A_1 F_4, G_2 C_3$	$A_1, 1, -, -$
E_8	$A_1, B_2, A_1 A_2, G_2 F_4$	$1, -, -, -$

In Section 7 we also define the labelled diagram of an arbitrary closed connected simple subgroup X of G, and prove the following result, which comes close to showing that the labelled diagram determines X up to $(\operatorname{Aut} G)$-conjugacy.

Theorem 8 *Let X_1, X_2 be closed connected simple subgroups of G of the same type, and of rank at least 2. Assume that X_1 and X_2 have the same labelled diagram, and that $p = 0$ or $p > N(X_1, G)$. Then with two exceptions, there is a subsystem subgroup Y of G such that*

(i) Y contains X_1 and a conjugate X_2^g of X_2,

(ii) Y is minimal subject to containing X_1, and

(iii) X_1 and X_2^g are conjugate in $\operatorname{Aut} Y$.

In the exceptions, $G = E_7$ (resp. E_8), $X_1 \cong X_2 \cong A_2$ and X_1, X_2 lie in subsystem subgroups A_5 and A_5' (resp. $A_2 A_5$ and $D_4 D_4$).

In the last sentence of Theorem 8, A_5 and A_5' are representatives of the two conjugacy classes of subsystem subgroups of type A_5 in E_7.

Notation

We use the following notation throughout the paper. If X is a connected reductive group over the algebraically closed field K, and λ is a dominant weight, then $V_X(\lambda)$ denotes the rational irreducible KX-module with high weight λ, and $W_X(\lambda)$ denotes the corresponding Weyl module. If $V_1, \ldots, V_k$ are modules, then $V_1 / \ldots / V_k$ denotes a module having the same composition factors as $V_1 \oplus \ldots \oplus V_k$. We often abbreviate this notation slightly as follows: if $\mu_1, \ldots, \mu_k$ are dominant weights for the group X, and $c_1, \ldots, c_k$ are positive integers, then

$$\mu_1^{c_1} / \ldots / \mu_k^{c_k}.$$

always denotes a KX-module having the same composition factors as $W_X(\mu_1)^{c_1} \oplus \ldots \oplus W_X(\mu_k)^{c_k}$.

When X is a subgroup of G and V is a KG-module, we use $V \downarrow X$ for the restriction of V to X. If the characteristic $p > 0$, and q is a power of p, then σ_q denotes a standard Frobenius morphism of X - that is, a morphism inducing the map $x_\alpha(t) \to x_\alpha(t^q)$ on root groups (notation of [St]). If V is a KX-module, we write $V^{(q)}$ for the KX-module obtained from V by twisting the action of X by σ_q (i.e. changing the action from $v \to vx$ to $v \to vx^{\sigma_q}$).

We use T_i to denote a torus of rank i, and $W(G)$ for the Weyl group of G. The fundamental roots in a fundamental system for G are denoted $\alpha_1, \ldots, \alpha_l$, and the corresponding fundamental dominant weights are $\lambda_1, \ldots, \lambda_l$. The Dynkin diagram of G is labelled as shown below, where the darkened nodes represent short roots.

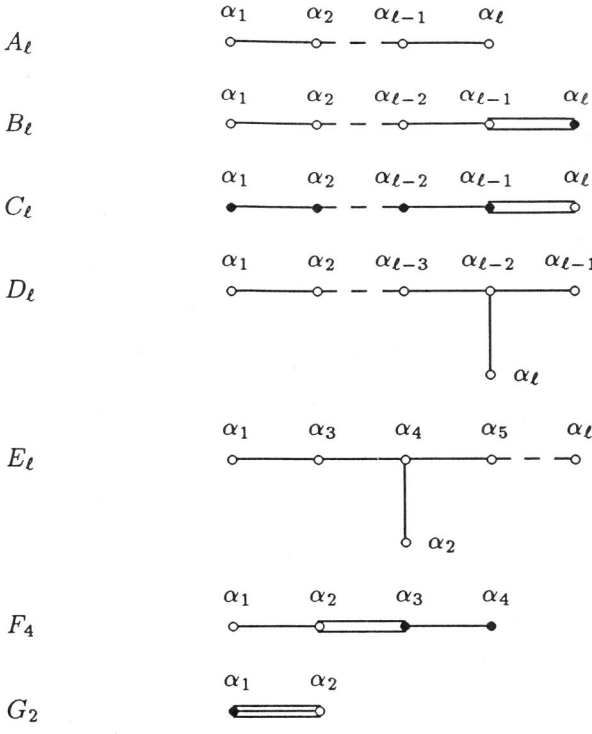

1 Preliminaries

In this section we develop our basic machinery for showing that for a large number of reductive groups X and irreducible X-modules V, the semidirect product XV has just one conjugacy class of closed complements to V. As outlined in the Introduction, this lies at the heart of the proof of Theorem 1. A key result is Proposition 1.5, which says that with one specific exception, if there is more than one class of closed complements then there is a rational indecomposable X-module which is an extension of V by the trivial module. (This is of course elementary if we drop the words "closed" and "rational".) We go on to apply this result to some particular examples of X and V.

Four preliminary lemmas are required to prove Proposition 1.5. The first two are well known to experts.

Lemma 1.1 *Let X be a classical connected simple algebraic group in characteristic p, with natural module $V = W_X(\lambda_1)$. Regard X as a subgroup of $GL(V)$, and $L(X)$ as an X-submodule of $V \otimes V^*$ ($\cong gl(V)$).*

> *(i) If $X = A_n$, then $L(X)$ is of codimension 1 in $V \otimes V^*$, and $Z(L(X))$ has dimension $\delta_{p,n+1}$.*

> *(ii) If $p \neq 2$ and $X = B_n$ or D_n, then $L(X) = \bigwedge^2 V$.*

> *(iii) If $p \neq 2$ and $X = C_n$, then $L(X) = S^2 V$.*

Lemma 1.2 *Let $\phi : X \to Y$ be a homomorphism of connected algebraic groups, with X simple. Assume that the differential $d\phi = 0$. Then ϕ can be factored through a Frobenius morphism of X.*

Proof. We may suppose that ϕ is surjective, and view both X and Y as Chevalley groups. Let K be the underlying algebraically closed field, of characteristic p. If U is a root subgroup of X, then U is isomorphic to K^+, with coordinate ring $K[x]$. Similarly, the image has coordinate ring $K[y]$. Using the fact that ϕ is a homomorphism and $d\phi = 0$, one easily checks that $\phi^*(y) = cx^q$, for some power q of p. A similar conclusion holds when ϕ is restricted to a 1-dimensional torus of X.

Fix a maximal torus T of X, and a corresponding system of root groups. Using the above remarks, together with the connectivity of the Dynkin diagram of X and the commutator relations, we see that for each root α in the root system of X, there is a power q_α of p and an element $k_\alpha \in K^*$ such that $\phi(U_\alpha(c)) = V_{\alpha'}(k_\alpha c^{q_\alpha})$, where $U_\alpha, V_{\alpha'}$ are root subgroups of X, Y, respectively. Conjugating by a suitable element of $\phi(T)$ we may re-parametrize root groups so that $k_\alpha = 1$ for all fundamental roots α. From the usual expressions for $h_\alpha(c)$ as products of elements of U_α and $U_{-\alpha}$, we find that $k_{-\alpha} = 1$ for each fundamental root α.

There is a Frobenius morphism $\delta : Y \to Y$ such that $\delta(V_{\alpha'}(d)) = V_{\alpha'}(d^p)$ for all fundamental roots α, and such that δ induces the p-power map on $A = \phi(T)$. Set $\gamma = \delta^{-1}\phi$. Then $\gamma : X \to Y$ is a well-defined homomorphism of abstract groups. Similarly, let F be the Frobenius morphism of X which induces the p-power map on T and on all T-root groups corresponding to fundamental roots and their negatives. Then ϕ and γF agree on a generating set for X. Hence $\phi = \gamma F$. It remains to show that γ is a morphism.

Write $X = \bigcup BwB$, the Bruhat decomposition (where each $w \in W(X)$, the Weyl group, is identified with a coset representative in X). For each w we have $BwB = ww^{-1}BwB \subseteq wU^-B$. Thus X is covered by a finite number of translates of the open dense set $C = U^-B$.

We claim that it suffices to show that $\gamma \downarrow C$ is a morphism. To see this, suppose $\gamma \downarrow C$ is a morphism and fix $g \in X$. Then $\gamma \downarrow gC = \lambda_{\phi(g)} \circ (\phi \downarrow C) \circ \lambda_{g^{-1}}$, where $\lambda_{g^{-1}} : X \to X$ is left multiplication by g^{-1} and $\lambda_{\phi(g)} : Y \to Y$ is left multiplication by $\phi(g)$. Each term in the composition is a morphism, hence so is $\gamma \downarrow gC$. It follows that there are principal open sets $U_{h_1}, \ldots, U_{h_n}$ covering X, such that $\gamma \downarrow U_{h_i}$ is a morphism for each i. Let f be in the coordinate ring of Y, and consider $\gamma^*(f)$. For each i there is an element g_i in the coordinate ring of X, and an integer n_i, such that $\gamma^*(f) \downarrow U_{h_i} = g_i/h_i^{n_i}$. As $U_{h_i^{n_i}} = U_{h_i}$, and since the open sets cover X, we conclude that the ideal generated by the functions $h_i^{n_i}$ ($1 \leq i \leq n$) must be the full coordinate ring of X. Hence there are elements $l_1, \ldots, l_n$ of the coordinate ring such that $1 = \sum l_i h_i^{n_i}$. Set $U = \bigcap U_{h_i}$. Then we have $\gamma^*(f) \downarrow U = \sum \gamma^*(f) l_i h_i^{n_i} = \sum l_i g_i$. Now U is a nonempty open set, so is dense in each U_{h_i}. It follows that $\gamma^*(f) = \sum l_i g_i$ on each U_{h_i}. Hence $\gamma^*(f) = \sum l_i g_i$ on X. This shows that γ is a morphism. Therefore it suffices to show that $\gamma \downarrow C$ is a morphism.

Now $C = U^-B = U^-TU$, and as a variety, $U^-TU = \prod U_{-\alpha} \times \prod T_i \times \prod U_\alpha$, where the first and last products are over all positive roots and the middle product is the direct product of 1-dimensional tori, the number of them being equal to the rank of X. The discussion of the first paragraph shows that γ restricted to each root group is a morphism. It remains to show that $\gamma \downarrow T$ is a morphism, where $T = \prod T_i$.

Write $K[T] = K[x_1^\pm, \ldots, x_n^\pm]$, where $K[T_i] = K[x_i^\pm]$ for each i. Similarly, let $\phi(T) = A = \prod A_i$, with $K[A] = K[y_1^\pm, \ldots, y_n^\pm]$. For each i, let δ_i, μ_i be isomorphisms from K^* to T_i, A_i, respectively. Then $\phi(\delta_i(a)) = \prod \mu_j(a^{c_{ij}})$ for integers c_{ij}. So for each k, we have $\phi^*(y_k)(\delta_i(a)) = y_k(\prod \mu_j(a^{c_{ij}})) = a^{c_{ik}}$. Hence $\phi^*(y_k) = \prod x_i^{c_{ik}}$. But then the hypothesis $d\phi = 0$ implies that p divides c_{ik} for all i, k. Writing $c_{ij} = pr_{ij}$, we have $\phi(\delta_i(a)) = \delta(\prod \mu_j(a^{r_{ij}}))$, and so $\gamma(\delta_i(a)) = \prod \mu_j(a^{r_{ij}})$, and as above $\gamma^*(y_k) = \prod x_i^{r_{ik}}$. Therefore $\gamma \downarrow T$ is a morphism, as required. $\square$

Lemma 1.3 *Let X be a simple connected algebraic group in characteristic $p > 0$, let q be a power of p, and let λ be a dominant weight for X. Suppose that there is a rational indecomposable extension of the irreducible module $V_X(q\lambda)$ by the trivial X-module. Then one of the following holds:*

(i) there is a rational indecomposable extension of $V_X(\lambda)$ by the trivial module;

(ii) $p = 2$, $X = C_n$ and $\lambda = 2^i\lambda_1$ for some i (or $2^i\lambda_2$ if $n = 2$).

Proof. Let W be a rational indecomposable extension $V_X(q\lambda)/V_X(0)$, and let $\phi : X \to SL(W)$ be the associated morphism. The differential $d\phi$ maps $L(X)$ to a nilpotent subalgebra of $sl(W)$.

Suppose that $d\phi \neq 0$. Then $L(X)$ has a nontrivial nilpotent quotient. Except when $X = C_n$, $p = 2$, it is easy to check that each root element of $L(X)$ is a commutator, so $L(X)$ has no nontrivial nilpotent quotient. Thus $X = C_n$ and $p = 2$. The ideal I of $L(X)$ generated by the short root elements and a Cartan subalgebra of $L(X)$ acts trivially on W; write $L(X) = I + \langle e_1, \ldots, e_{2n}\rangle$ and pick $w \in W - V_X(q\lambda)$. As X is trivial on $W/V_X(q\lambda)$, it follows that $\langle e_1 w, \ldots, e_{2n} w\rangle$ is X-invariant. Hence $\dim V_X(\lambda) \leq 2n$, which forces $\lambda = 2^i\lambda_1$ (or $2^i\lambda_2$ if $n = 2$), as in conclusion (ii).

Now suppose that $d\phi = 0$. Then by 1.2, $\phi = \psi\sigma_{q'}$, where $\sigma_{q'}$ is a q'-power Frobenius morphism of X and ψ is a morphism $X \to SL(W)$ with $d\psi \neq 0$. Moreover $q' \leq q$, and ψ corresponds to a rational indecomposable extension $V_X(\frac{q}{q'}\lambda)/V_X(0)$. If $q' < q$ then as above, (ii) holds. And if $q' = q$ then (i) holds. $\square$

For the next result, let $X = X_1 \ldots X_k$ be a commuting product of simple connected algebraic groups X_i in characteristic p. For each i, let V_i be a nontrivial rational irreducible finite-dimensional X_i-module, and set $V = V_1 \otimes \ldots \otimes V_k$, an irreducible X-module.

Lemma 1.4 *With the above notation, if there is a rational indecomposable extension of V by the trivial module, then the same holds for each V_i.*

Proof. Suppose W is such an extension of V. Now $V \downarrow X_i$ is a homogenenous direct sum of X_i-modules isomorphic to V_i. If there is no rational indecomposable extension of V_i by the trivial module, then $W \downarrow X_i$ is completely reducible, and so $C_W(X_i)$ is a 1-dimensional space. But $C_W(X_i)$ is X-invariant, contrary to the fact that W is indecomposable. $\square$

The next proposition is a key result. In it, we view the semidirect product XV as an algebraic group.

Proposition 1.5 *As above, let $X = X_1 \ldots X_k$, $V = V_1 \otimes \ldots \otimes V_k$, where for each i, $V_i = V_{X_i}(\mu_i)$. If the semidirect product XV contains more than one conjugacy class of closed complements to V, then for each i, one of the following holds:*

(i) there is a rational indecomposable extension of V_i by the trivial module;

(ii) $p = 2$, $X_i = C_n$ and $\mu_i = 2^j\lambda_1$ (or $2^j\lambda_2$ if $n = 2$).

Proof. We recall some standard material. A 1-cocycle is a (rational) function $\gamma : X \to V$ such that $\gamma(x_1 x_2) = \gamma(x_1) x_2 + \gamma(x_2)$ for all $x_1, x_2 \in X$. There is an action of X on the additive group of 1-cocycles, given by

$$\gamma^g(x) = \gamma(gx) - \gamma(g)$$

for all $g, x \in X$. For $v \in V$ there is a 1-coboundary $\bar{v}$ given by $\bar{v}(x) = v - vx$ for all $x \in X$. With the above action of X, we have $\bar{v}^g = \overline{vg}$ for $v \in V, g \in X$. And for a 1-cocycle γ and $g \in X$,

$$\gamma^g - \gamma = -\overline{\gamma(g)},$$

so X acts trivially on the factor group of 1-cocycles modulo 1-coboundaries, namely $H^1(X, V)$.

Now suppose that Y is a closed complement to V in XV, and that Y is not V-conjugate to X. Then

$$Y = \{x\gamma(x) : x \in X\}$$

where $\gamma : X \to V$ is a 1-cocycle. Write x' for the element $x\gamma(x)$ of Y.

Form the K-vector space $W = V + \langle \gamma \rangle$, and extend the action of X on V to an action on W by setting

$$\gamma x = \gamma - \gamma(x)$$

for $x \in X$. This action also extends to Y via the sequence $Y \to XV \to X \to GL(W)$ (where the first map is inclusion and the second is projection).

We claim that the representation $Y \to GL(W)$ is rational. The action of X on V is rational, so it is clear from the above sequence that the action of Y on V is also rational. Now consider the sequence $Y \to XV \to V$, where the first map is inclusion and the second is projection. Both maps are morphisms, so the map $x' \to \gamma(x)$ is rational. The claim follows.

Suppose that Y fixes a vector in $W - V$. Then there exists $v \in V$ such that $(v + \gamma)x' = v + \gamma$ for all $x' \in Y$. It follows that $(v + \gamma)x = v + \gamma$ for all $x \in X$, so by definition of γx,

$$\gamma(x) = -(v - vx)$$

for all $x \in X$. Hence $\gamma = -\bar{v}$, which means that X and Y are conjugate by $-v$, a contradiction. Therefore W is indecomposable under the action of Y.

Consider again the map $Y \to XV \to X$, and write $Y = Y_1 \ldots Y_k$, where for each i, Y_i is the preimage of X_i. This gives a morphism $\phi_i : Y_i \to X_i$ which is an isomorphism of abstract groups. We then get a morphism $Y_i \to X_i \to GL(V_i)$, and Lemma 1.4 shows that as a Y_i-module, V_i has a rational indecomposable extension W_i by the trivial module.

Suppose now that $L(Y_i)$ has no nonzero nilpotent ideal. Then $L(Y_i) \cap L(V) = 0$. Since the differential of the projection map $X_i V \to X_i$ has kernel $L(V)$, it follows that $d\phi_i$ is an isomorphism. Hence ϕ_i is an isomorphism of algebraic groups by [Sp,

4.3.4]. Now the map $X_i \to Y_i \to GL(W_i)$ (where the first map is ϕ_i^{-1}) shows that conclusion (i) holds.

Thus we may assume that $L(Y_i)$ has a nonzero nilpotent ideal I which is the kernel of $d\phi_i$. It follows that $Y_i = B_n$, $p = 2$ and I is the ideal generated by short root elements of $L(Y_i)$; and the image of $L(Y_i)$ under $d\phi_i$ is an ideal in $L(X_i)$, whence $X_i = C_n$. We claim that Y_i is of adjoint type. For if not, then Y_i is simply connected and the ideal generated by short root elements of $L(Y_i)$ contains $Z(L(Y_i))$ (since $Z(L(Y_i))$ lies in the Lie algebra of a short SL_2 in Y_i). But $Z(L(Y_i))$ is not nilpotent, which is a contradiction.

Hence Y_i is adjoint, and so there is a surjective morphism $\tau : X_i \to Y_i$. Then $\tau\phi_i$ is a Frobenius morphism σ_q of X_i, so it follows that there is a rational indecomposable extension of the X_i-module $V_i^{(q)}$ by the trivial module. The result now follows from 1.3. $\square$

The following standard result enables us to use Proposition 1.5 to show that for many of the semidirect products XV as in the hypothesis (of 1.5), there is just one class of closed complements to V. Recall that $W_X(\lambda)$ denotes the Weyl module for X with high weight λ, and $\mathrm{Ext}_X^1(V, K)$ denotes the group of equivalence classes of rational extensions of V by the trivial X-module K.

Proposition 1.6 [Ja, p.207] *Let X be a simple algebraic group over K, and λ a dominant weight. Then*

$$Ext_X^1(V_X(\lambda), K) \cong Hom_X(rad(W_X(\lambda)), K).$$

Corollary 1.7 *Let X be a simple algebraic group over K of characteristic p, and let $V = V_X(\lambda)$, where if $(X, p) = (C_n, 2)$ then $\lambda \neq 2^j\lambda_1$ (or $2^j\lambda_2$ if $n = 2$). Suppose that $Hom_X(rad(W_X(\lambda)), K) = 0$. Then the semidirect product XV contains just one class of closed complements to V.*

Proof. This is immediate from 1.5 and 1.6. $\square$

In the next few results, we present some information on Weyl modules which is designed for the application of 1.5 and 1.7. Throughout, X is a simple algebraic group in characteristic p.

In the first proposition, for a positive integer r we label by $V_X(r)$ the irreducible module for $X = A_1$ with high weight $r\lambda_1$.

Proposition 1.8 *Let $X = A_1$, and suppose that there is a rational indecomposable extension of $V_X(r)$ by the trivial module. Then $p > 0$ and $V_X(r)$ is a Frobenius twist of the module*

$$V_X(p - 2) \otimes V_X(1)^{(p)}.$$

Proof. This follows from [AJL, 3.9]. $\square$

Proposition 1.9 *Let $X = A_2$, and let $a, b \in \{0, 1, \ldots, p - 1\}$.*

   (i) If $a + b + 2 \leq p$ then $W_X(a\lambda_1 + b\lambda_2)$ is irreducible.

   (ii) If $W_X(a\lambda_1 + b\lambda_2)$ is reducible, then it is indecomposable with two composition factors, $V_X(a\lambda_1 + b\lambda_2)$ and $V_X((p - b - 2)\lambda_1 + (p - a - 2)\lambda_2)$.

   (iii) $W_X(a\lambda_1 + b\lambda_2)$ has dimension $\frac{1}{2}(a + 1)(b + 1)(a + b + 2)$.

Proof. Part (iii) is the Weyl degree formula, and (i) and (ii) are easy applications of the sum formula in [And]. □

A similar application of [And] yields the next result.

Proposition 1.10 *Let $X = B_2$ and $a, b \in \{0, \ldots, p - 1\}$.*

   (i) If $2a + b + 3 \leq p$, or if $a = 0$, then $W_X(a\lambda_1 + b\lambda_2)$ is irreducible.

   (ii) $W_X(a\lambda_1)$ is irreducible unless $\frac{1}{2}(p - 3) < a \leq p - 3$, in which case $W_X(a\lambda_1)$ has two composition factors, $V_X(a\lambda_1)$ and $V_X((p - a - 3)\lambda_1)$.

   (iii) $W_X(a\lambda_1 + b\lambda_2)$ has dimension $\frac{1}{6}(a + 1)(b + 1)(a + b + 2)(2a + b + 3)$.

Proposition 1.11 *Let X, λ be one of the following:*

$$
\begin{aligned}
X = A_r \ (r \geq 3): \quad & \lambda = c\lambda_1, \lambda_2 \text{ or } \lambda_3 (0 \leq c \leq p - 1) \\
X = B_r \ (r \geq 2): \quad & \lambda = \lambda_1 (p \neq 2) \text{ or } \lambda_r \\
X = D_r \ (r \geq 4): \quad & \lambda = \lambda_1, \lambda_{r-1} \text{ or } \lambda_r \\
X = E_6, E_7: \quad & \lambda = \lambda_1, \lambda_7 \ (resp.)
\end{aligned}
$$

Then $W_X(\lambda)$ is irreducible.

Proof. Except for the case where $(X, \lambda) = (A_r, c\lambda_1)$ or (B_r, λ_1), the result is immediate since the Weyl group of X is transitive on the weights of $W_X(\lambda)$. The case $(A_r, c\lambda_1)$ follows from [Se1, 1.14]; and for the other case, $W_X(\lambda_1)$ is the natural module for B_r, which is irreducible. □

Proposition 1.12 *Let X, λ be one of the following:*

X	λ
G_2	$\lambda_1 \ (p \neq 2)$, λ_2, $2\lambda_1 \ (p \neq 2, 7)$, $\lambda_1 + \lambda_2 \ (p \neq 3, 7)$
A_3	$\lambda_1 + \lambda_3 \ (p \neq 2)$, $\lambda_1 + \lambda_2$
$B_3 \ (p \neq 2)$	λ_2, $2\lambda_3$, $\lambda_1 + \lambda_3$
$C_3 \ (p \neq 2)$	λ_1, $\lambda_2 \ (p \neq 3)$, λ_3, $\lambda_1 + \lambda_2$
$C_4 \ (p \neq 2)$	λ_1, λ_2, λ_3
D_4	λ_1, λ_3, λ_4, $\lambda_2 \ (p \neq 2)$, $\lambda_1 + \lambda_3 \ (p \neq 2)$, $\lambda_1 + \lambda_4 \ (p \neq 2)$, $\lambda_3 + \lambda_4 \ (p \neq 2)$
F_4	$\lambda_4 \ (p \neq 3)$

Then $W_X(\lambda)$ has no trivial composition factors.

Proof. All the modules listed occur in the tables in [BW], from which the result is immediate. □

The next lemma gives us another tool for obtaining a conclusion like that of 1.7.

Lemma 1.13 *Let X be a simple algebraic group and let V be a finite-dimensional rational irreducible X-module. Suppose that X contains tori $E_1, \ldots, E_r$ satisfying the following conditions:*

1. *each E_i is fixed-point-free on $V - \{0\}$,*

2. *$[E_i, E_j] = 1$ for all i, j,*

3. *$X = \langle C_X(E_i) : 1 \leq i \leq r \rangle$.*

Then the semidirect product XV contains just one class of closed complements to V. In particular, this holds if $Z(X)$ acts trivially on V.

Proof (cf. [LS2,1.8]). Let X_0 be a closed complement to V in XV. We claim that E_1 lies in a unique conjugate of X_0 in XV. For X_0 contains a torus F_1 which is a complement to V in $E_1 V$. Then $E_1 = F_1^v$ for some $v \in V$, so $E_1 \leq X_0^v$. If also $E_1 \leq X_0^w$ ($w \in V$), then $F_1^{vw^{-1}} \leq X_0 \cap F_1 V = F_1$. Hence $vw^{-1} \in N_V(F_1) = 1$ (since F_1 is fixed-point-free), so $v = w$, proving the claim.

Replacing X_0 by a suitable V-conjugate, we may therefore assume that $E_1 \leq X_0$. Since E_1 is fixed-point-free, $C_{XV}(E_1) = C_{X_0 V}(E_1) \leq X_0$. Hence for any i, $E_i \leq C_X(E_1) \leq X_0$. Then $C_X(E_i) \leq X_0$, and so $X = \langle C_X(E_i) : 1 \leq i \leq r \rangle \leq X_0$. Therefore $X_0 = X$, giving the result. □

In the following proposition, we adopt the notation $ab\ldots$ to denote the irreducible module $V_X(a\lambda_1 + b\lambda_2 + \ldots)$.

Proposition 1.14 *Let X be one of the following simple algebraic groups in characteristic p, and let V be one of the irreducible X-modules listed, where q denotes a positive power of p:*

X	V
$A_2 \ (p > 3)$	$10 \otimes 10^{(q)}, 10 \otimes 01^{(q)}, 10 \otimes 20^{(q)}, 10 \otimes 02^{(q)},$ $10 \otimes 11^{(q)}, 20 \otimes 10^{(q)}, 02 \otimes 10^{(q)}, 11 \otimes 10^{(q)}$
$B_2 \ (p \neq 2)$	$11, 13, 01 \otimes 01^{(q)}, 01 \otimes 10^{(q)},$ $01 \otimes 02^{(q)}, 10 \otimes 01^{(q)}, 02 \otimes 01^{(q)}$
$A_3 \ (p > 2)$	$100 \otimes 100^{(q)}, 100 \otimes 010^{(q)}, 100 \otimes 001^{(q)},$ $010 \otimes 100^{(q)}, 001 \otimes 100^{(q)}$
B_3	$001 \otimes 001^{(q)}$

Then there is just one class of closed complements to V in XV, and the same holds for XV^* (where V^* is the dual of V).

Proof. Consider $X = A_2$. When V or V^* is $10 \otimes 11^{(q)}$ or $11 \otimes 10^{(q)}$, $Z(X)$ acts nontrivially, so the result holds. For the other cases we apply 1.13 with $r = 2$ and

$$E_1 = \{\mathrm{diag}\,(\alpha, \alpha, \alpha^{-2}) : \alpha \in K^*\}, \ E_2 = \{\mathrm{diag}\,(\alpha^{-2}, \alpha, \alpha) : \alpha \in K^*\}.$$

Next let $X = B_2$. If V is $11, 13, 01 \otimes 10^{(q)}, 10 \otimes 01^{(q)}, 01 \otimes 02^{(q)}$ or $02 \otimes 01^{(q)}$, then $Z(X)$ acts nontrivially. And if $V = 01 \otimes 01^{(q)}$, regard X as $C_2 = Sp_4$, and relative to a symplectic basis for the usual Sp_4-module, take $r = 2$ and

$$E_1 = \{\mathrm{diag}\,(\alpha, \alpha, \alpha^{-1}, \alpha^{-1}) : \alpha \in K^*\}, \ E_2 = \{\mathrm{diag}\,(\alpha, \alpha^{-1}, \alpha^{-1}, \alpha) : \alpha \in K^*\}.$$

Then 1.13 applies.

For all cases with $X = A_3$, we let $r = 2$ and take E_1, E_2 as above.

Finally, consider $X = B_3$ and $V = 001 \otimes 001^{(q)}$. Write W for the 8-dimensional irreducible X-module 001. Now X contains a subgroup $D = D_3$ preserving a decomposition of W as a direct sum of two 4-spaces W_1 and W_2. Relative to a basis of W containing bases of W_1 and W_2, D contains a subgroup

$$E_1 = \{\mathrm{diag}\,(\alpha, \alpha, \alpha^{-1}, \alpha^{-1}, \alpha^{-1}, \alpha^{-1}, \alpha, \alpha) : \alpha \in K^*\}.$$

On the usual 7-dimensional module for X, this subgroup E_1 acts as $\{\mathrm{diag}\,(\alpha^2, \alpha^{-2}, 1, 1, 1, 1, 1) : \alpha \in K^*\}$, so $C_X(E_1)$ contains B_2. Therefore, if we define

$$E_2 = \{\mathrm{diag}\,(\alpha, \alpha^{-1}, \alpha, \alpha^{-1}, \alpha^{-1}, \alpha, \alpha^{-1}, \alpha) : \alpha \in K^*\}$$

(relative to the above basis of W), then $\langle C_X(E_1), C_X(E_2) \rangle = X$ and hence 1.13 again gives the result. $\square$

Proposition 1.15 *Suppose that either* $X = B_2, V = 10 \otimes 10^{(q)}$ *and* $p > 2$, *or* $X = G_2, V = 10 \otimes 10^{(q)}$ *and* $p > 3$ *(where* $q = p^a, a \geq 1$*). Then the conclusion of* 1.14 *holds.*

Proof. First consider $X = B_2, V = 10 \otimes 10^{(q)}$. Let X_0 be a closed complement to V in XV. If A is a subgroup A_1 of X generated by a short root group and its opposite, then A acts as SO_3 on $V_X(10)$, so has composition factors of high weights $2, 0, 0$. Hence the nontrivial composition factors of A on V are $2 \otimes 2^{(q)}, 2$ and $2^{(q)}$. By 1.8, none of these has a rational indecomposable extension by the trivial module, so by 1.5, there is only one class of closed complements to V in AV. Therefore, conjugating X_0 by an element of V, we have $A \leq X \cap X_0$.

Now $C_{XV}(A)^0 = T_1 V_0$, where V_0 is a 4-dimensional subspace of V and T_1 is a 1-dimensional torus in X. Conjugating X by an element of $C_{XV}(A)$, we may assume that $AT_1 \leq X \cap X_0$. Then $X \cap X_0$ contains a maximal torus of X, so there is a 1-dimensional torus T_1' in $X \cap X_0$ such that T_1' lies in a fundamental A_1 in X. Then $C_{XV}(T_1')^0 = V_1 A' T_1'$, where V_1 is a 1-space in V and A' is a fundamental A_1 in X.

Since A' is the unique SL_2 in $C_{XV}(T_1')^0$, we deduce that $A' \leq X \cap X_0$. Thus $X \cap X_0$ contains $\langle AT_1, A' \rangle$, which is equal to X, proving the result in this case.

Now let $X = G_2, V = 10 \otimes 10^{(q)}$. Again let X_0 be a closed complement to V in XV. Pick a fundamental subgroup $A \cong A_1$ in X. The nontrivial composition factors of A on V are $1 \otimes 1^{(q)}, 1$ and $1^{(q)}$. Since $p > 3$, none of these extends the trivial module by 1.8, so as above we may assume that $A \leq X \cap X_0$. Also $C_{XV}(A)^0 = V_0 A'$, where $V_0 = C_V(A)$ and $A' = C_X(A) \cong A_1$. Moreover V_0 is the irreducible A'-module $2 \otimes 2^{(q)}$, so again using 1.8 and 1.5, we may take $AA' \leq X \cap X_0$.

Let T_2 be a maximal torus in AA'. Then $C_{XV}(T_2)^0 = V_1 T_2$, where $V_1 = C_V(T_2)$ is a 1-space; and there is an element $w \in N_X(T_2)$ of order 3 which centralizes V_1. Thus $N_{XV}(T_2)$ contains a unique subgroup isomorphic to $T_2 \langle w \rangle$, from which it follows that $T_2 \langle w \rangle \leq X \cap X_0$. Therefore $X \cap X_0$ contains $\langle AA', w \rangle$, which is equal to X. This completes the proof. $\square$

Proposition 1.16 *Let $X < Y$ be simple algebraic groups in characteristic p, with $(X, Y, p) = (B_n, D_{n+1}, 2) \, (n \geq 2)$, $(F_4, E_6, 3)$ or $(C_3, A_5, 3)$, in each case the natural embedding. Let $\lambda = \lambda_1, \lambda_1$ or λ_2 in the respective cases, and $V = V_Y(\lambda)$. Then there is just one class of closed complements to V in XV.*

Proof. In each case there is an X-composition series $0 < V_1 < V_2 < V$ with V_1 and V/V_2 both trivial 1-dimensional X-modules; moreover, V is indecomposable for X (see [LS2,1.6]). Write $W = V/V_1$, and suppose that there is more than one class of closed complements in XW. The proof of 1.5 implies that either there is a rational indecomposable extension of W by the trivial X-module, or $X = C_n$ and $p = 2$. The latter is not the case, by hypothesis. Hence W extends the trivial module. Writing $V_2/V_1 = V_X(\mu)$, 1.6 now implies that $\operatorname{Hom}_X(\operatorname{rad}(W_X(\mu)), K)$ has dimension at least 2, which is a contradiction. $\square$

2 Some restrictions and tensor products

Let G be a simple algebraic group in characteristic p, and X a connected reductive subgroup of G. In this section we prove a number of results which give the composition factors of the restriction $V \downarrow X$ for various G-modules V. We also give the composition factors of some tensor products for groups of small rank.

We recall some notation from the Introduction. If $V_1, \ldots, V_k$ are X-modules, then $V_1 / \ldots / V_k$ denotes an X-module having the same composition factors as $V_1 \oplus \ldots \oplus V_k$. And if X is simple and λ is a dominant weight, we often write just λ to denote a module with the same composition factors as the Weyl module $W_X(\lambda)$. Combining these notations, if $\mu_1, \ldots, \mu_k$ are dominant weights, then $\mu_1 / \ldots / \mu_k$ denotes a module with the same composition factors as $W_X(\mu_1) \oplus \ldots \oplus W_X(\mu_k)$.

Proposition 2.1 *Let G be F_4, E_6, E_7 or E_8. The following table gives the composition factors of the restriction of $L(G)/L(M)$ to M, for various closed connected subgroups M of maximal rank in G.*

G	M	M-comp. factors of $L(G)/L(M)$
F_4	B_4	λ_4
	$A_1 C_3$	$1 \otimes \lambda_3$
	$A_2 A_2 \ (p \neq 2)$	$\lambda_1 \otimes 2\lambda_2 / \lambda_2 \otimes 2\lambda_1$
E_6	$A_1 A_5$	$1 \otimes \lambda_3$
	$A_2 A_2 A_2$	$\lambda_1 \otimes \lambda_1 \otimes \lambda_1 / \lambda_2 \otimes \lambda_2 \otimes \lambda_2$
E_7	A_7	λ_4
	$A_1 D_6$	$1 \otimes \lambda_5$
	$A_2 A_5$	$\lambda_1 \otimes \lambda_2 / \lambda_2 \otimes \lambda_4$
E_8	A_8	λ_3 / λ_6
	D_8	λ_7
	$A_1 E_7$	$1 \otimes \lambda_7$
	$A_2 E_6$	$\lambda_1 \otimes \lambda_6 / \lambda_2 \otimes \lambda_1$
	$D_4 D_4$	$\lambda_1 \otimes \lambda_1 / \lambda_3 \otimes \lambda_3 / \lambda_4 \otimes \lambda_4$
	$A_4 A_4$	$\lambda_1 \otimes \lambda_2 / \lambda_2 \otimes \lambda_4 / \lambda_3 \otimes \lambda_1 / \lambda_4 \otimes \lambda_3$

Proof. This follows from [Se2, 1.8] and its proof. (The prime restrictions assumed in [Se2, 1.8] are used only to express $L(G) \downarrow M$ as a direct sum, and are not needed to give the composition factors.) $\square$

Proposition 2.2 *The restriction of $L(E_8)$ to its maximal rank subgroup A_2^4 is as follows:*

$$
\begin{aligned}
L(E_8) \downarrow A_2^4 \ = \ & L(A_2)^4 / \lambda_1 \otimes \lambda_1 \otimes \lambda_1 \otimes 0 / \lambda_2 \otimes \lambda_2 \otimes \lambda_2 \otimes 0 / \lambda_1 \otimes \lambda_2 \otimes 0 \otimes \lambda_2 / \\
& \lambda_2 \otimes \lambda_1 \otimes 0 \otimes \lambda_1 / \lambda_1 \otimes 0 \otimes \lambda_2 \otimes \lambda_1 / \lambda_2 \otimes 0 \otimes \lambda_1 \otimes \lambda_2 / \\
& 0 \otimes \lambda_1 \otimes \lambda_2 \otimes \lambda_2 / 0 \otimes \lambda_2 \otimes \lambda_1 \otimes \lambda_1.
\end{aligned}
$$

Proof. Write $X = A_2^4 = X_1 X_2 X_3 X_4$ with $X_i \cong A_2$. Then $X \le EX_4 \cong E_6 A_2$, and by 2.1, $L(E_8) \downarrow EX_4 = L(E)/L(X_4)/\lambda_1 \otimes \lambda_2/\lambda_6 \otimes \lambda_1$.

Consider now the restriction of the 27-dimensional module $V_E(\lambda_6)$ to $X_1 X_2 X_3$, where we take the latter group to have fundamental system $\alpha_1, \alpha_3, \alpha_2, \alpha_0 - \alpha_2, \alpha_6, \alpha_5$ in the E_6 root system. A maximal vector v^+ affords the weight λ_6 for E, and so affords the weight $0 \otimes \lambda_2 \otimes \lambda_1$ for $X_1 X_2 X_3$. For an E_6-root α, denote by f_α the usual element of $L(E_6)$; and if $\alpha = a_1\alpha_1 + \ldots + a_6\alpha_6$, write $\alpha = a_1 a_2 a_3 a_4 a_5 a_6$. Then one checks that $f_{000111}v^+$ and $f_{011211}v^+$ are also maximal vectors for $X_1 X_2 X_3$. These afford the E_6-weights $\lambda_6 - 000111$ and $\lambda_6 - 011211$, hence afford the weights $\lambda_2 \otimes \lambda_1 \otimes 0$ and $\lambda_1 \otimes 0 \otimes \lambda_2$ for $X_1 X_2 X_3$, respectively. Thus the composition factor $\lambda_6 \otimes \lambda_1$ for EX_4 gives factors $0 \otimes \lambda_2 \otimes \lambda_1 \otimes \lambda_1/\lambda_2 \otimes \lambda_1 \otimes 0 \otimes \lambda_1/\lambda_1 \otimes 0 \otimes \lambda_2 \otimes \lambda_1$ for $X = X_1 X_2 X_3 X_4$. As $L(E_8)$ is self-dual, the duals of these also appear in $L(E_8) \downarrow X$. Finally, consider $L(E) \downarrow X_1 X_2 X_3$. The maximal vector e_{112221} affords the weight $\lambda_2 \otimes \lambda_2 \otimes \lambda_2$. The dual also appears, giving all the factors in the conclusion. $\square$

Proposition 2.3 *Let V_{56} be the 56-dimensional module $V_{E_7}(\lambda_7)$ and let V_{27} be the 27-dimensional module $V_{E_6}(\lambda_1)$. The restrictions of V_{56} and V_{27} to various subsystem subgroups of E_7 and E_6 are as follows:*

$$V_{56} \downarrow A_7 = \lambda_2/\lambda_6,$$
$$V_{56} \downarrow A_1 D_6 = 1 \otimes \lambda_1/0 \otimes \lambda_5,$$
$$V_{56} \downarrow A_2 A_5 = \lambda_1 \otimes \lambda_1/\lambda_2 \otimes \lambda_5/0 \otimes \lambda_3,$$
$$V_{56} \downarrow E_6 = \lambda_1/\lambda_6/0^2,$$
$$V_{27} \downarrow A_1 A_5 = 1 \otimes \lambda_1/0 \otimes \lambda_4,$$
$$V_{27} \downarrow A_2 A_2 A_2 = \lambda_1 \otimes \lambda_2 \otimes 0/\lambda_2 \otimes 0 \otimes \lambda_1/0 \otimes \lambda_1 \otimes \lambda_2,$$
$$V_{27} \downarrow D_4 = \lambda_1/\lambda_3/\lambda_4/0^3.$$

Proof. First consider $V_{56} \downarrow A_7$, where the A_7 has fundamental system $\alpha_1, \alpha_3, \alpha_4, \alpha_5, \alpha_6, \alpha_7, 1223210$ (with the labelling of the E_7 root system as in the Introduction). To avoid confusion, label the fundamental dominant weights for the A_7 as $\lambda_1', \ldots, \lambda_7'$ relative to the above ordering. A maximal vector in V_{56} affords λ_6' for A_7, so $V_{56} \downarrow A_7$ has a composition factor $V_{A_7}(\lambda_6')$; and since V_{56} is self-dual, there is also a factor $V_{A_7}(\lambda_2')$.

Now consider $V_{56} \downarrow A_1 D_6$, where the $A_1 D_6$ has fundamental system $\alpha_0, \alpha_3, \alpha_2, \alpha_4, \alpha_5, \alpha_6, \alpha_7$. Label the fundamental dominant weights for D_6 relative to this ordering as $\lambda_6', \lambda_5', \ldots, \lambda_1'$. By considering a maximal vector v^+ in V_{56}, we see that $V_{56} \downarrow A_1 D_6$ has a composition factor $1 \otimes \lambda_1'$. And the vector $f_{1011111}v^+$ (notation as in the proof of 2.2) is a maximal vector for $A_1 D_6$ affording weight $0 \otimes \lambda_5'$. Therefore $V_{56} \downarrow A_1 D_6 = 1 \otimes \lambda_1'/0 \otimes \lambda_5'$.

Similar considerations handle $V_{56} \downarrow A_2 A_5, V_{56} \downarrow E_6, V_{27} \downarrow A_1 A_5$ and $V_{27} \downarrow D_4$, while the result for $V_{27} \downarrow A_2 A_2 A_2$ follows from the proof of 2.2. $\square$

In the next two results we give the restrictions $L(G) \downarrow X$ and $V_{56} \downarrow X, V_{27} \downarrow X$, where X is a maximal closed connected subgroup of G not containing a maximal torus, as given in [Se2, Theorem 1].

Proposition 2.4 *Let G be an exceptional simple algebraic group in characteristic p, and let X be a maximal closed connected subgroup of G not containing a maximal torus, as given in [Se2, Theorem 1]. Then the X-composition factors of $L(G)/L(X)$ are as follows:*

G	X	X-comp. factors of $L(G)/L(X)$
G_2	A_1 $(p = 0$ or $p \geq 7)$	10
F_4	$A_1 G_2$ $(p \neq 2)$	$4 \otimes 10$
	G_2 $(p = 7)$	11
	A_1 $(p = 0$ or $p \geq 13)$	$22/14/10$
E_6	$A_2 G_2$	$11 \otimes 10$
	C_4 $(p \neq 2)$	0001
	F_4	0001
	G_2 $(p \neq 2, 3, 7)$	11
	A_2 $(p \neq 2, 3)$	$41/14$
E_7	$G_2 C_3$	$10 \otimes 010$
	$A_1 F_4$	$2 \otimes 0001$
	$A_1 G_2$ $(p \neq 2)$	$4 \otimes 10/2 \otimes 20$
	$A_1 A_1$ $(p \neq 2, 3)$	$2 \otimes 8/4 \otimes 6/6 \otimes 4/2 \otimes 4/4 \otimes 2$
	A_2 $(p \neq 2, 3, 5)$	44
	A_1 $(p = 0$ or $p \geq 17)$	$26/22/18/16/14/10^2/6$
	A_1 $(p = 0$ or $p \geq 19)$	$34/26/22/18/14/10$
E_8	$G_2 F_4$	$10 \otimes 0001$
	$A_1 A_2$ $(p \neq 2, 3)$	$6 \otimes 11/4 \otimes 30/4 \otimes 03/2 \otimes 22$
	B_2 $(p \neq 2, 3, 5)$	$06/32$
	A_1 $(p = 0$ or $p \geq 23)$	$38/34/28/26/22^2/18/16/14/10/6$
	A_1 $(p = 0$ or $p \geq 29)$	$46/38/34/28/26/22/18/14/10$
	A_1 $(p = 0$ or $p \geq 31)$	$58/46/38/34/26/22/14$

Proof. See [Se2, p.193]. □

Proposition 2.5 *Let $G = E_6$ or E_7 and let V_{56}, V_{27} be as in 2.3.*

(i) If X is as in 2.4, then $V_{27} \downarrow X$ and $V_{56} \downarrow X$ are as in the table on the following page.

(ii) If $G = E_6$ and X is a maximal A_1 in C_4 or F_4 in G, then $V_{27} \downarrow X = 12/8/4$ or $16/8/0$, respectively.

Proof. First consider $G = E_7$. If $G = G_2 C_3$, then the factor G_2 lies in a subsystem subgroup D_4, and the factor C_3 in a factor A_5 of a maximal rank subgroup $A_2 A_5$

G	X	$V_{27} \downarrow X$	$V_{56} \downarrow X$
E_6	$A_2 G_2$	$10 \otimes 10/02 \otimes 00$	
	$C_4 \, (p \neq 2)$	0100	
	F_4	$0001/0000$	
	$G_2 \, (p \neq 2,3,7)$	20	
	$A_2 \, (p \neq 2,3)$	22	
E_7	$G_2 C_3$		$10 \otimes 100/00 \otimes 001$
	$A_1 F_4$		$1 \otimes 0001/3 \otimes 0000$
	$A_1 G_2 \, (p \neq 2)$		$1 \otimes 01/3 \otimes 10$
	$A_1 A_1 \, (p \neq 2,3)$		$6 \otimes 3/4 \otimes 1/2 \otimes 5$
	$A_2 \, (p \neq 2,3,5)$		$60/06$
	$A_1 \, (p = 0 \text{ or } p \geq 17)$		$21/15/11/5$
	$A_1 \, (p = 0 \text{ or } p \geq 19)$		$27/17/9$

(see the proof of [Se2, 3.16]). Using 2.3, we see that

$$V_{56} \downarrow C_3 = 100^7/001, \quad V_{56} \downarrow G_2 = 10^6/00^{14}.$$

Since $W_{C_3}(001)$ has dimension 14, it follows that $V_{56} \downarrow G_2 C_3 = 10 \otimes 100/00 \otimes 001$.

Now let $X = A_1 F_4$. This subgroup is constructed in [Se2, 3.12], and the factor F_4 lies in a subsystem subgroup E_6 of G. The restriction $V_{56} \downarrow E_6$ is given in 2.3, and it is easy to see that $V_{E_6}(\lambda_i) \downarrow F_4 = \lambda_4/0$ for $i = 1, 6$. Hence

$$V_{56} \downarrow F_4 = \lambda_4^2/0^4.$$

By [Se2, 3.12], the A_1 factor of X has labelled diagram

$$0 \quad 0 \quad 0 \quad 0 \quad 0 \quad 2$$
$$0$$

(This means that if $T_1 = \{T(c) : c \in K^*\}$ is a maximal torus of the A_1, then $T(c)e_{\alpha_7} = c^2 e_{\alpha_7}$ and $T(c)e_{\alpha_i} = e_{\alpha_i}$ for $i \leq 6$.) Since

$$\lambda_7 = \frac{1}{2}(2\alpha_1 + 3\alpha_2 + 4\alpha_3 + 6\alpha_4 + 5\alpha_5 + 4\alpha_6 + 3\alpha_7),$$

the labelled diagram given above means that the highest weight of T_1 on V_{56} is 3. Applying all the other weights in the $W(E_7)$-orbit of λ_7 to T_1, we see that the weights of T_1 on V_{56} are $\pm 3, \pm 1^{27}$. Hence

$$V_{56} \downarrow A_1 = 3/1^{26}.$$

It follows that $V_{56} \downarrow A_1 F_4 = 1 \otimes \lambda_4/3 \otimes 0$.

Next consider $X = A_1 G_2 (p \neq 2)$. By [Se2, p.34], the G_2 factor lies in a subsystem group A_6, which in turn lies in a subgroup A_7. The restriction $V_{56} \downarrow A_7$ is given in 2.3, from which we deduce that

$$V_{56} \downarrow A_6 = \lambda_1 / \lambda_6 / \lambda_2 / \lambda_5.$$

We have $V_{A_6}(\lambda_i) \downarrow G_2 = 10$ for $i = 1, 6$; and by 2.10 below, $V_{A_6}(\lambda_i) \downarrow G_2 = 10/01$ for $i = 2, 5$. Hence

$$V_{56} \downarrow G_2 = 10^4 / 01^2.$$

As for the A_1 factor of X, by [Se2, 3.12] its labelled diagram has a 2 labelling the α_2 node, and 0 on the other nodes. Applying the weights in the $W(E_7)$-orbit of λ_7 as above, we find that $V_{56} \downarrow A_1 = 3^7 / 1^{14}$. Hence $V_{56} \downarrow A_1 G_2 = 1 \otimes 01 / 3 \otimes 10$.

When $X = A_1 A_1 (p \neq 2, 3)$, the labelled diagrams for the two factors A_1 have their only labels 2 on the α_4 and α_5 nodes, respectively. Arguing as before, we see that V_{56} restricts to the A_1 factors as $6^4 / 4^2 / 2^6$ and $5^3 / 3^7 / 1^5$, respectively. Hence $V_{56} \downarrow A_1 A_1 = 6 \otimes 3 / 4 \otimes 1 / 2 \otimes 5$.

Now let $X = A_2 (p \neq 2, 3, 5)$. This maximal subgroup of E_7 is constructed in [Se2, 5.8], which gives explicitly the Lie algebra $L(X) = \langle e_\alpha, f_\alpha, e_\beta, f_\beta \rangle$ (see [Se2, p.89]). If $P_8 = QL$ is the standard E_7-parabolic of E_8, then $Z(Q) = U_{\alpha_0}$ (where α_0 is the longest root in the E_8 root system), and $V_{56} \cong Q/Z(Q)$ as L'-modules. Modulo $Z(Q)$, the vector $u_{\alpha_0 - \alpha_8}(1)$ is a maximal vector for L', and we calculate that

$$u_{\alpha_0 - \alpha_8}(1)^{h_\beta(k)} = u_{\alpha_0 - \alpha_8}(1), \quad u_{\alpha_0 - \alpha_8}(1)^{h_\alpha(k)} = u_{\alpha_0 - \alpha_8}(k^6).$$

Hence $u_{\alpha_0 - \alpha_8}(1)$ is a maximal vector for X affording high weight 60. It follows that $V_{56} \downarrow X$ has composition factors 60 and 06. These both have dimension 28 (see 1.9), so the result follows in this case.

To complete the proof for $G = E_7$, suppose $X = A_1$. By [Se2, 4.1] there are two conjugacy classes of maximal connected subgroups A_1 in E_7, with labelled diagrams

$$\begin{array}{ccccccc} 2 & 2 & 0 & 2 & 2 & 2 \\ & & 2 & & & \end{array} \quad \text{and} \quad \begin{array}{cccccc} 2 & 2 & 2 & 2 & 2 & 2 \\ & & & 2 & & \end{array}$$

As above, we apply the weights in the $W(E_7)$-orbit of λ_7 to obtain the conclusion.

Now let $G = E_6$. First consider $X = A_2 G_2$, as constructed in [Se2, 3.15]. The factor G_2 lies in a subsystem group D_4, so we see from 2.3 that

$$V_{27} \downarrow G_2 = 10^3 / 00^6.$$

Let A be the factor A_2 of X. Since it centralizes a subgroup A_2 of the G_2 factor, A lies in a subsystem group A_2^3, and projects trivially to one of the factors (modulo $Z(A_2^3)$). For a fixed copy D of A_2, write the elements of this A_2^3 as $a.b.c$ ($a, b, c \in D$). Taking A_2^3 to act on $L(E_6)$ as in 2.1, and using the fact that $L(E_6) \downarrow A = 11^7$ by

2.4, we can assume that $A = \{a^\tau.a.1 : a \in D\}$, where τ is a graph automorphism of D. Therefore 2.3 gives

$$V_{27} \downarrow A = 01 \otimes 01/10^6 = 02/10^7.$$

It follows that $V_{27} \downarrow A_2 G_2 = 10 \otimes 10/02 \otimes 00$, as required.

It is well known and easy to see that $V_{27} \downarrow F_4 = \lambda_4/0$. Finally, when X is $C_4(p \neq 2)$, $G_2(p \neq 2, 3, 7)$ or $A_2(p \neq 2, 3)$, $V_{27} \downarrow X$ is irreducible with high weight 0100, 20 or 22 respectively, by [Te]. $\square$

We now turn our attention to restrictions of various modules for classical groups.

Proposition 2.6 *Let $Y = D_n$ and let V_{n-1}, V_n be the spin modules $V_Y(\lambda_{n-1})$, $V_Y(\lambda_n)$, respectively. Define X_{n-1}, X_n to be the Levi subgroups A_{n-1} of Y corresponding to the subsystems $\langle\alpha_1, \ldots, \alpha_{n-1}\rangle, \langle\alpha_1, \ldots, \alpha_{n-2}, \alpha_n\rangle$, respectively.*

(i) If n is odd, then as A_{n-1}-modules,

$$V_{n-1} \downarrow X_{n-1},\ V_n \downarrow X_n,\ (V_n \downarrow X_{n-1})^*,\ (V_{n-1} \downarrow X_n)^* = \lambda_{n-1}/\lambda_{n-3}/\ldots/\lambda_4/\lambda_2/0.$$

(ii) If n is even, then

$$V_{n-1} \downarrow X_{n-1},\ V_n \downarrow X_n = \lambda_{n-1}/\lambda_{n-3}/\ldots/\lambda_3/\lambda_1,$$

$$V_n \downarrow X_{n-1},\ V_{n-1} \downarrow X_n = \lambda_{n-2}/\lambda_{n-4}/\ldots/\lambda_4/\lambda_2/0^2.$$

Proof. Suppose n is odd, and set $V = V_{n-1}$. Let Q be the unipotent radical of a parabolic subgroup with Levi factor X_{n-1}. We make repeated use of [Se1, 2.1 and 2.3]. By [Se1, 2.1], $V/[Q, V]$ is irreducible for X_{n-1} with high weight λ_{n-1}. Next, note that V has a weight $\lambda = \lambda_{n-1} - \alpha_{n-2} - \alpha_{n-1} - \alpha_n$ (in the $W(D_n)$-orbit of λ_{n-1}), and this restricts to X_{n-1} as the A_{n-1}-weight λ_{n-3}. Hence by [Se1, 2.3], $[Q, V]/[Q, Q, V]$ has an X_{n-1}-composition factor with high weight λ_{n-3}. Arguing similarly with the weight $\lambda_{n-1} - \alpha_{n-4} - 2\alpha_{n-3} - 3\alpha_{n-2} - 2\alpha_{n-1} - 2\alpha_n$ of V, we see that $[Q, Q, V]/[Q, Q, Q, V]$ has an X_{n-1}-composition factor with high weight λ_{n-5}. Repeating, we see that $V \downarrow X_{n-1}$ has composition factors with high weights $\lambda_{n-1}, \lambda_{n-3}, \ldots, \lambda_4, \lambda_2, 0$. The sum of the dimensions of these factors is $2^{n-1} = \dim V$, so

$$V \downarrow X_{n-1} = \lambda_{n-1}/\lambda_{n-3}/\ldots/\lambda_2/0.$$

As n is odd, the long word $w_0 \in W(D_n)$ interchanges λ_{n-1} with $-\lambda_n$, so $V_{n-1}^* \cong V_Y(-w_0(\lambda_{n-1})) = V_n$. And X_{n-1} and X_n are interchanged by a graph automorphism of Y, which also interchanges λ_{n-1} and λ_n, so $V_{n-1} \downarrow X_{n-1}$, $V_n \downarrow X_n$ have the same composition factors.

The proof of part (i) is now complete. Part (ii) is similar and is left to the reader. $\square$

Proposition 2.7 *Let $Y = B_n (n \geq 3)$ or $D_{n+1} (n \geq 4)$, and let X be either a Levi subgroup $B_r (r \geq 1)$ or $D_r (r \geq 3)$ of Y, or a subgroup B_n of $Y = D_{n+1}$. If V is a spin module $V_Y(\lambda_n)$ (or $V_Y(\lambda_{n+1})$ if $Y = D_{n+1}$), then all composition factors of $V \downarrow X$ are spin modules for X.*

Proof. When X is a maximal Levi subgroup, we use the argument of the proof of 2.6; then the general Levi subgroup case follows by induction. And when $X = B_n < Y = D_{n+1}$, a maximal vector for Y in V affords the weight λ_n for X; since $\dim V = \dim V_X(\lambda_n) = 2^n$, we conclude that $V \downarrow X = V_X(\lambda_n)$, a spin module. $\square$

The next two results classify various low-dimensional modules which we shall work with later in detail. As always, p denotes the underlying characteristic, and q is a power of p.

Proposition 2.8 *Let X be a simple algebraic group of rank at least 2, and suppose that $V = V_X(\lambda)$ is such that $\dim V \leq 16$ and V is self-dual if $\dim V > 9$. Then X and λ are as follows; we give λ up to field twists and duals, and we also give $Cl(V)$, the smallest classical group on V containing X (except in a few cases with $p = 2$).*

X	λ	$Cl(V)$
A_n, B_n, C_n, D_n	λ_1	X (C_n if $(X,p) = (B_n, 2)$)
A_2	$20 \, (p \neq 2)$	A_5
	$\lambda_i \otimes \lambda_j^{(q)}$	A_8
	11	D_4 (B_3 if $p = 3$)
B_2	$20 \, (p \neq 2)$	D_7 (B_6 if $p = 5$)
	$02 \, (p \neq 2)$	D_5
	11	$C_8 \, (p \neq 2)$
	$01 \otimes 01^{(q)}$	D_8
	$\lambda_i \otimes \lambda_j^{(q)} \, (p = 2)$	D_8
G_2	10	B_3 (C_3 if $p = 2$)
	01	D_7 (B_3 if $p = 3$)
A_3	010	D_3
	101	$B_7 \, (p \neq 2)$
B_3	001	D_4
	$010 \, (p = 2)$	D_7
C_3	010	D_7 (B_6 if $p = 3$)
	001	$C_7 \, (p \neq 2), D_4 \, (p = 2)$
B_4	0001	D_8

Proof. For $X = A_2, B_2$, we use 1.9 and 1.10 to find the possibilities for λ. For other groups X, we count $W(X)$-conjugates of λ and of subdominant weights which are known to occur as weights of V (for example using [Te, 1.30]). The total number of such conjugates is at most 16, which forces λ to be as given, apart from a few possibilities which are ruled out using the tables in [BW].

As for the group $Cl(V)$, this is $SL(V)$ if V is not self-dual; and if V is self-dual and $p \neq 2$, $Cl(V)$ is determined using [St, Lemma 79]. The assertions for $p = 2$ can be found in [Se1, Table 1]. $\square$

Proposition 2.9 *Let $X = A_1$ and let V be a rational irreducible X-module such that $\dim V \leq 16$, and $X \leq SO(V)$ if $\dim V > 9$. Then up to field twists, V is as follows, where r denotes the irreducible X-module of hight weight r, and $Cl(V)$ is the smallest classical group containing X.*

V	p	$Cl(V)$
r	$p = 0$ or $p > r$	Sp_{r+1} (r odd), SO_{r+1} (r even)
$1 \otimes 1^{(q)}$	any	D_2
$1 \otimes 2^{(q)}, 2 \otimes 1^{(q)}$	$p \neq 2$	C_3
$1 \otimes 3^{(q)}, 3 \otimes 1^{(q)}$	$p \neq 2,3$	D_4
$1 \otimes 1^{(q)} \otimes 1^{(q')}$ ($q \neq q'$)	any	C_4 (D_4 if $p = 2$)
$2 \otimes 2^{(q)}$	$p \neq 2$	B_4
$1 \otimes 5^{(q)}, 5 \otimes 1^{(q)}$	$p \neq 2,3,5$	D_6
$1 \otimes 1^{(q)} \otimes 2^{(q')}, 2 \otimes 1^{(q)} \otimes 1^{(q')}$	$p \neq 2$	D_6
$2 \otimes 4^{(q)}, 4 \otimes 2^{(q)}$	$p \neq 2,3$	B_7
$1 \otimes 7^{(q)}, 7 \otimes 1^{(q)}$	$p \neq 2,3,5,7$	D_8
$3 \otimes 3^{(q)}$	$p \neq 2,3$	D_8
$1 \otimes 1^{(q)} \otimes 1^{(q')} \otimes 1^{(q'')}$	any	D_8

Proof. This is a routine exercise. The group $Cl(V)$ is determined using the fact that the form on V preserved by X is the product of the forms on the tensor factors. $\square$

For many of the embeddings $X \leq Cl(V)$ given by 2.8 and 2.9, we shall need to know the restrictions to X of the $Cl(V)$-modules $\wedge^2 V$, $\wedge^3 V$, and also of spin modules for $Cl(V)$. These restrictions are given in the next few results.

Proposition 2.10 *Let X, λ be as in 2.8, and $V = V_X(\lambda)$; assume also that $X \leq SO(V)$ if $\dim V > 9$. In the following table, we list the high weights of the X-composition factors of $\wedge^2 V$, and also of $\wedge^3 V$ when $\dim V \leq 9$ (assuming some extra restrictions on p, as stated).*

X	λ	comp. factors of $\bigwedge^2 V$	comp. factors of $\bigwedge^3 V$
A_n	λ_1	λ_2	λ_3
$B_n\ (p \neq 2)$	λ_1	$\lambda_2\ (n \geq 3),\ 2\lambda_2\ (n = 2)$	$\lambda_3\ (n \geq 4),\ 2\lambda_3\ (n = 3),$ $2\lambda_2\ (n = 2)$
$C_n\ (p \neq 2)$	λ_1	$\lambda_2,\ 0$	$\lambda_3,\ \lambda_1$
$D_n\ (p \neq 2, n \geq 4)$	λ_1	λ_2	$\lambda_3\ (n \geq 5),$ $\lambda_3 + \lambda_4\ (n = 4)$
$A_2\ (p \neq 2, 3)$	20	21	$30, 03$
	$11\ (p \neq 5)$	$30, 03, 11$	$22, 30, 03, 11, 00$
$B_2\ (p \neq 2)$	$20\ (p \neq 5)$	$22, 02$	
	02	$12, 02$	
	$01 \otimes 01^{(q)}$	$02, 02^{(q)}, 10 \otimes 02^{(q)},$ $02 \otimes 10^{(q)}$	
$G_2\ (p \neq 2)$	10	$10, 01$	$20, 10, 00$
	$01\ (p \neq 3)$	$30, 01$	
A_3	$010\ (p \neq 2)$	101	$200, 002$
	$101\ (p \neq 2)$	$210, 012, 101$	
$B_3\ (p \neq 2)$	001	$100, 010$	$101, 001$
$C_3\ (p \neq 2, 3)$	010	$200, 101$	
B_4	0001	$0100, 0010$	

Proof. The method for finding the composition factors of $\bigwedge^2 V$ and $\bigwedge^3 V$ is straightforward: we calculate a suitable set of weights of V; then any sum of two distinct weights is a weight of $\bigwedge^2 V$, and any sum of three is a weight of $\bigwedge^3 V$. Using this, together with the dimensions of irreducible X-modules given by [BW] (and by 1.9, 1.10), we obtain the conclusion. As an example, we carry out this calculation for $X = A_2, \lambda = 11\ (p \neq 2, 3, 5)$.

The weights of $V_{A_2}(11)$ are 11, 2 -1, -12, 00 and their negatives. Therefore $\bigwedge^2 V$ has composition factors with high weights 30 and 03. The 11-weight space in $\bigwedge^2 V$ has dimension 3, and the 11-weight spaces in $V_X(30)$ and $V_X(03)$ have dimension 1. Hence $\bigwedge^2 V$ also has a composition factor $V_X(11)$. By 1.9, the sum of the dimensions of $V_X(30), V_X(03)$ and $V_X(11)$ is 28, so these are all the composition factors of $\bigwedge^2 V$.

Now consider $\bigwedge^3 V$ with $V = V_{A_2}(11)$. The highest weight occurring in $\bigwedge^3 V$ is 22, so $V_X(22)$ is a composition factor, of dimension 27. Observe that the 30-weight space in $\bigwedge^3 V$ has dimension 2, while that in $V_X(22)$ has dimension 1 (by Freudenthal's formula, [Hu, p.122]). Hence $V_X(30)$, and similarly $V_X(03)$, are composition factors of $\bigwedge^3 V$. The 11-weight-spaces in $\bigwedge^3 V, V_X(22), V_X(30), V_X(03)$ are respectively of dimensions 5,2,1,1. Consequently $\bigwedge^3 V$ has a composition factor $V_X(11)$. Similarly we see that $\bigwedge^3 V$ has a trivial composition factor. The dimensions of the composition factors found so far add up to 56, so we have found them all. $\square$

The same method of proof gives the next result.

Proposition 2.11 *Let* $X = A_2(p \neq 2,3,5)$ *or* $C_4(p \neq 2)$, *and let* V *be* $V_{A_2}(11)$ *or* $V_{C_4}(1000)$, *respectively. Then*

$$\bigwedge^4 V = 22^2/11^2 \ \text{if} \ X = A_2,$$
$$\bigwedge^4 V = 0001/0100/0000 \ \text{if} \ X = C_4.$$

Proposition 2.12 *Let* $X, \lambda, Cl(V)$ *be as in 2.8, and write* $Y = Cl(V)$. *Suppose that* $Y = B_n$ *or* D_n *(with* $n \geq 5$ *if* $Y = D_n$*) and* $X < Y$. *If* W *is a spin module* $V_Y(\lambda_n)$ *(or* $V_Y(\lambda_{n-1})$ *if* $Y = D_n$*), then the high weights of the composition factors of* $W \downarrow X$ *are as follows:*

Embedding $X < Y$	λ	comp. factors of $W \downarrow X$
$B_2 < D_5 \ (p \neq 2)$	02	11 (and 01 if $p = 5$)
$B_2 < D_7 \ (p \neq 2,3,5)$	20	13 (and 03 if $p = 7$)
$B_2 < D_8 \ (p \neq 2,5)$	$01 \otimes 01^{(q)}$	$20, 20^{(q)}, 10 \otimes 02^{(q)}, 02 \otimes 10^{(q)}$ or
		$01 \otimes 11^{(q)}, 11 \otimes 01^{(q)}$
$G_2 < B_3 \ (p \neq 2)$	10	10, 00
$G_2 < D_7 \ (p \neq 3,7)$	01	11
$A_3 < B_7 \ (p \neq 2)$	101	111 (and 200, 002 if $p = 3$, 010 if $p = 5$)
$C_3 < D_7 \ (p \neq 2,3)$	010	110 (and 100 if $p = 7$)
$B_4 < D_8 \ (p \neq 2)$	0001	1001 (and 0001 if $p = 3$) or
		2000, 0010 (and 0000 if $p = 3$)

Proof. For $(X,Y) = (G_2, B_3)$, the spin module W is 8-dimensional, and it is clear that $W \downarrow X = 10/00$.

Next observe that if $Y = D_n$ with n odd, then $V_Y(\lambda_n)$ and $V_Y(\lambda_{n-1})$ are dual to each other, so their restrictions to X have the same composition factors (as these composition factors must be self-dual by 2.8).

For $(X,Y) = (B_2, D_5), (B_2, D_7), (G_2, D_7)$ or (C_3, D_7), it follows from [Se1, Table 1] that $W \downarrow X$ is the Weyl module for X with high weight 11, 13, 11 or 110, respectively, except possibly when this Weyl module is reducible. In the latter case $W \downarrow X$ has the same weights as the corresponding Weyl module, hence has the same composition factors, which can be read off from [BW] when $X = C_3$, and using the sum formula in [And] when $X = B_2$. (Alternatively, the embeddings considered in this paragraph can be handled using the method given for $(X,Y) = (A_3, B_7)$ below.)

It remains to deal with $(X, Y) = (B_2, D_8), (A_3, B_7)$ or (B_4, D_8). Consider first $X = B_2, Y = D_8, \lambda = 01 \otimes 01^{(q)}$. Here $X < C_2 \otimes C_2 < Y$, and we shall see in the proof of the next proposition (2.13) that

$$
\begin{aligned}
V_Y(\lambda_7) \downarrow C_2 C_2 &= 10 \otimes 11/11 \otimes 10 \\
V_Y(\lambda_8) \downarrow C_2 C_2 &= 20 \otimes 01/01 \otimes 20/02 \otimes 00/00 \otimes 02.
\end{aligned}
$$

The result follows for this case.

Next consider $X = A_3, Y = B_7$. Let $\alpha_i \, (1 \leq i \leq 3)$ and $\beta_i \, (1 \leq i \leq 7)$ be fundamental roots for X and Y, respectively. The module $V_Y(\lambda_1) \downarrow X$ is isomorphic to $L(X)$, which we take to have basis ordered as

$$
e_{\alpha_1 + \alpha_2 + \alpha_3}, e_{\alpha_1 + \alpha_2}, e_{\alpha_2 + \alpha_3}, e_{\alpha_1}, e_{\alpha_2}, e_{\alpha_3}, h_{\alpha_1}, h_{\alpha_2}, h_{\alpha_3}, e_{-\alpha_3}, \ldots, e_{-\alpha_1 - \alpha_2 - \alpha_3}.
$$

Relative to this basis, we calculate that for $t \in K$,

$$
\begin{aligned}
h_{\alpha_1}(t) &= \operatorname{diag}(t, t, t^{-1}, t^2, t^{-1}, 1, 1, 1, 1, 1, t, t^{-2}, t, t^{-1}, t^{-1}) \\
&= h_{\beta_1}(t) h_{\beta_2}(t^2) h_{\beta_3}(t) h_{\beta_4}(t^3) h_{\beta_5}(t^2) h_{\beta_6}(t^2) h_{\beta_7}(t).
\end{aligned}
$$

Similarly,

$$
\begin{aligned}
h_{\alpha_2}(t) &= h_{\beta_2}(t) h_{\beta_3}(t^2) h_{\beta_4}(t) h_{\beta_5}(t^3) h_{\beta_6}(t^2) h_{\beta_7}(t), \\
h_{\alpha_3}(t) &= h_{\beta_1}(t) h_{\beta_3}(t) h_{\beta_4}(t) h_{\beta_6}(t^2) h_{\beta_7}(t).
\end{aligned}
$$

Thus both the weights λ_7 and $\lambda_7 - \beta_7$ of $W = V_Y(\lambda_7)$ afford the weight 111 on the maximal torus $\langle h_{\alpha_i}(t) : 1 \leq i \leq 3, t \in K^* \rangle$ of X. It follows that $W \downarrow X$ has two composition factors $V_X(111)$. For $p \neq 3, 5$, $V_X(111) = W_X(111)$ has dimension 64 (see [BW]), so $W \downarrow X$ has only the two composition factors $V_X(111)$; and for $p = 3, 5$, $W \downarrow X$ has the same weights as $W_X(111) \oplus W_X(111)$, so has the same composition factors, which are as given in the conclusion (see [BW]).

Finally, let $(X, Y) = (B_4, D_8)$. Write $W_1 = V_Y(\lambda_7)$, $W_2 = V_Y(\lambda_8)$. The actions of Y on W_1 and W_2 are those of the two half-spin groups of type D_8; one of these, say Y^{W_1}, contains $X = \operatorname{Spin}_9$, and the other, Y^{W_2}, contains $X = SO_9$. From [Se1, Table 1], we see that $W_1 \downarrow X = V_X(1001)$ if $p \neq 3$, and $W \downarrow X$ has the same composition factors as $W_X(1001)$ if $p = 3$, namely $V_X(1001)$ and $V_X(0001)$. Now consider $W_2 \downarrow X$, with $X = SO_9$. Take a Levi subgroup $B = B_3$ of X. Then B lies in a Levi subgroup A_7 of Y, where $W_2 \downarrow A_7 = \lambda_2/\lambda_4/\lambda_6/0^2$ by 2.6. Therefore $W_2 \downarrow B = 100^3/010^2/200/002/000^3$. It follows that all composition factors of $W_2 \downarrow X$ are of the form $V_X(\lambda)$ with $\lambda = ab00, a0c0$ or $a00d$ for some non-negative integers a, b, c, d satisfying $b \leq 2, c \leq 1, d \leq 2$. If $d = 1$ then $Z(X) \neq 1$, a contradiction; and if $d = 2$ then the restriction of $V_X(a002)$ to B gives two composition factors 002 (use [Se1, 2.1]), which conflicts with the composition factors of $W_2 \downarrow B$ given above. Hence $d = 0$. It is now an easy exercise to check that the only possible composition factors of $W_2 \downarrow X$ of dimension at most 124 are $V_X(\lambda)$ with $\lambda = 1000, 0100, 1100, 2000, 0010$ or 0000.

Pick a subsystem subgroup $D = D_4$ of X. Then $Z(D) \cong Z_2$. Moreover, D is a diagonal subgroup of a subsystem subgroup $D_4 D_4 \leq Y$, and by 2.7, $W_2 \downarrow D_4 D_4$ has two composition factors, both of the form $\lambda_i \otimes \lambda_j$ ($i,j = 1, 3$ or 4). If $i \neq j$ then the restriction to D of this factor has a composition factor $V_D(\lambda_i + \lambda_j)$ of dimension 56 (see [BW]); and $V_D(\lambda_i) \otimes V_D(\lambda_i)$ has composition factors $V_D(2\lambda_i)$, $V_D(\lambda_2)$ and $V_D(0)$. Consequently $V_X(1100)$ does not appear in $W_2 \downarrow X$. Now the only possible factor $V_X(\lambda)$ whose restriction to B has $V_B(002)$ is $V_X(0010)$, of dimension 84; so this must appear in $W_2 \downarrow X$. And to obtain $V_B(200)$, the module $V_X(2000)$, of dimension 44 (43 if $p = 3$), must appear. Since $84 + 44 = 128 = \dim W_2$, this establishes the result for this case. $\square$

Proposition 2.13 *Let $X = A_1$ and let V, $Cl(V)$ be as in the table of 2.9. Suppose that W is either $\bigwedge^2 V$, a spin module for $Cl(V)$, or $\bigwedge^3 V$ with $\dim V \leq 9$. Then $W \downarrow X$ is as in the table on the following page.*

Proof. The composition factors of $\bigwedge^2 V$ and $\bigwedge^3 V$ are routine to find, in the manner of the proof of 2.10.

Now let W be a spin module for $Cl(V)$. Write $V = V_X(r)$; then r is even, say $r = 2s$, and $Cl(V) = B_s$. Moreover the spin module W has high weight $\lambda = \frac{1}{2}(\alpha_1 + 2\alpha_2 + \ldots + s\alpha_s)$ (where $\alpha_1, \ldots \alpha_s$ are fundamental roots for $Cl(V)$). Since X is a regular A_1 in $Cl(V)$, λ affords the weight $1 + 2 + \ldots + s$ on a maximal torus T_X of X, and this is the highest weight of a composition factor of $W \downarrow X$. The next highest weights for T_X are afforded by the following weights of W: $\lambda - \alpha_s$, $\lambda - \alpha_{s-1} - \alpha_s$, $\lambda - \alpha_{s-2} - \alpha_{s-1} - \alpha_s$ and $\lambda - \alpha_{s-1} - 2\alpha_s$. Therefore the T_X-weight $(1 + 2 + \ldots + s) - 6$ occurs twice in W (and is the highest such), and so $W \downarrow X$ has a composition factor with this high weight. Continuing in this fashion, we obtain all the composition factors of $W \downarrow X$.

Now consider the remaining possibilities for V, namely $2 \otimes 2^{(q)}$, $5 \otimes 1^{(q)}$, $2 \otimes 1^{(q)} \otimes 1^{(q')}$, $4 \otimes 2^{(q)}$, $7 \otimes 1^{(q)}$, $3 \otimes 3^{(q)}$ and $1 \otimes 1^{(q)} \otimes 1^{(q')} \otimes 1^{(q'')}$, with $Cl(V) = B_4$, D_6, D_6, B_7, D_8, D_8 and D_8, respectively. Here $X < Y < Cl(V)$, where $Y = B_1 \otimes B_1$, $C_3 \otimes C_1$, $C_3 \otimes C_1$, $B_2 \otimes B_1$, $C_4 \otimes C_1$, $C_2 \otimes C_2$ or $C_1 \otimes C_1 \otimes C_1 \otimes C_1$, respectively. In each case we first determine the composition factors of the restriction $W \downarrow Y$. We handle just the last three cases in detail, the others being similar and easier.

First let $V = 7 \otimes 1^{(q)}$, so $X < Y = C_4 \otimes C_1 < D_8$. There are two spin modules to consider, namely $W_1 = V_{D_8}(\lambda_7)$ and $W_2 = V_{D_8}(\lambda_8)$. If C is the factor C_4 of Y, then C lies in a Levi subgroup A_7 of D_8, and by 2.6 we may assume that

$$W_1 \downarrow A_7 = \lambda_1/\lambda_3/\lambda_5/\lambda_7, \quad W_2 \downarrow A_7 = \lambda_2/\lambda_4/\lambda_6/0^2.$$

Then using 2.10 and 2.11 for $C_4 < A_7$, we have

$$W_1 \downarrow C = \lambda_1^4/\lambda_3^2, \quad W_2 \downarrow C = \lambda_2^3/\lambda_4/0^5.$$

Similarly, if D is the factor C_1 of Y, then D also lies in a Levi subgroup A_7, and

$$W_1 \downarrow D = 1^{48}/3^8, \quad W_2 \downarrow D = 4/2^{27}/0^{42}$$

V	$\bigwedge^2 V \downarrow X$	$\bigwedge^3 V \downarrow X$	$W \downarrow X$, W spin module
r	$2r-2/2r-6/\dots$	$3r-6/3r-10/\dots$	$1\,(r=2); 3\,(r=4); 6/0\,(r=6);$ $10/4\,(r=8); 15/9/5\,(r=10);$ $21/15/11/9/3\,(r=12);$ $28/\dots\,(r=14)$
$1\otimes 1^{(q)}$	$2/2^{(q)}$		
$2\otimes 1^{(q)}$	$2\otimes 2^{(q)}/4/0$	$4\otimes 1^{(q)}/2\otimes 1^{(q)}/3^{(q)}$	
$3\otimes 1^{(q)}$	$4\otimes 2^{(q)}/6/2^{(q)}/2$	$7\otimes 1^{(q)}/5\otimes 1^{(q)}/$ $3\otimes 3^{(q)}/3\otimes 1^{(q)}/$ $1\otimes 1^{(q)}$	
$1\otimes 1^{(q)}\otimes 1^{(q')}$	$2\otimes 2^{(q)}/2\otimes 2^{(q')}/$ $2^{(q)}\otimes 2^{(q')}/0$	$3\otimes 1^{(q)}\otimes 1^{(q')}/$ $1\otimes 3^{(q)}\otimes 1^{(q')}/$ $1\otimes 1^{(q)}\otimes 3^{(q')}/$ $1\otimes 1^{(q)}\otimes 1^{(q')}$	
$2\otimes 2^{(q)}$	$4\otimes 2^{(q)}/2\otimes 4^{(q)}/$ $2/2^{(q)}$	$4\otimes 4^{(q)}/4\otimes 2^{(q)}/$ $2\otimes 4^{(q)}/6/6^{(q)}/$ $2\otimes 2^{(q)}/2/2^{(q)})$	$3\otimes 1^{(q)}/1\otimes 3^{(q)}$
$5\otimes 1^{(q)}$	$8\otimes 2^{(q)}/4\otimes 2^{(q)}$ $2^{(q)}/10/6/2$		$5\otimes 2^{(q)}/9/3$ or $8\otimes 1^{(q)}/4\otimes 1^{(q)}/3^{(q)}$
$2\otimes 1^{(q)}\otimes 1^{(q')}$	$2\otimes 2^{(q)}\otimes 2^{(q')}/$ $4\otimes 2^{(q)}/4\otimes 2^{(q')}/$ $2/2^{(q)}/2^{(q')}$		$2\otimes 1^{(q)}\otimes 2^{(q')}/4\otimes 1^{(q)}/3^{(q)}$ or $2\otimes 2^{(q)}\otimes 1^{(q')}/4\otimes 1^{(q')}/3^{(q')}$
$4\otimes 2^{(q)}$	$6\otimes 4^{(q)}/8\otimes 2^{(q)}/$ $4\otimes 2^{(q)}/2\otimes 4^{(q)}/$ $6/2^{(q)}/2$		$(9/5/3)\otimes 1^{(q)}/(7/5/1)\otimes 3^{(q)}/$ $3\otimes 5^{(q)}$
$7\otimes 1^{(q)}$	$12\otimes 2^{(q)}/8\otimes 2^{(q)}/$ $4\otimes 2^{(q)}/2^{(q)}/$ $14/10/6/2$		$7\otimes 3^{(q)}/(15/11/9/5/3)\otimes 1^{(q)}$ or $(12/8/4)\otimes 2^{(q)}/4^{(q)}/16/10/8/4$
$3\otimes 3^{(q)}$	$6\otimes 4^{(q)}/4\otimes 6^{(q)}/$ $4\otimes 2^{(q)}/2\otimes 4^{(q)}/$ $6/6^{(q)}/2/2^{(q)}$		$(7/5/1)\otimes 3^{(q)}/3\otimes (7/5/1)^{(q)}$ or $(6/2)\otimes 4^{(q)}/4\otimes (6/2)^{(q)}/$ $8/4/8^{(q)}/4^{(q)}$
$1\otimes 1^{(q)}\otimes$ $1^{(q')}\otimes 1^{(q'')}$	$2\otimes 2^{(q)}\otimes 2^{(q')}/$ $2\otimes 2^{(q)}\otimes 2^{(q'')}/$ $2\otimes 2^{(q')}\otimes 2^{(q'')}/$ $2^{(q)}\otimes 2^{(q')}\otimes 2^{(q'')}/$ $2/2^{(q)}/2^{(q')}/2^{(q'')}$		$3\otimes 1^{(q)}\otimes 1^{(q')}\otimes 1^{(q'')}/$ $1\otimes 3^{(q)}\otimes 1^{(q')}\otimes 1^{(q'')}/$ $1\otimes 1^{(q)}\otimes 3^{(q')}\otimes 1^{(q'')}/$ $1\otimes 1^{(q)}\otimes 1^{(q')}\otimes 3^{(q'')}$ or $2\otimes 2^{(q)}\otimes 2^{(q')}/2\otimes 2^{(q)}\otimes 2^{(q'')}/$ $2\otimes 2^{(q')}\otimes 2^{(q'')}/2^{(q)}\otimes 2^{(q')}\otimes 2^{(q'')}/$ $4/4^{(q)}/4^{(q')}/4^{(q'')}$

(or vice versa). Hence

$$W_1 \downarrow C_4C_1 = \lambda_1 \otimes 3/\lambda_3 \otimes 1, \quad W_2 \downarrow C_4C_1 = \lambda_2 \otimes 2/\lambda_4 \otimes 0/0 \otimes 4.$$

So to determine $W_1 \downarrow X$ and $W_2 \downarrow X$ we must find the composition factors of the restrictions of $V_{C_4}(\lambda_i)$ $(i = 1,2,3,4)$ to a regular A_1 in C_4. Since $V_{C_4}(\lambda_i) \subseteq \bigwedge^i V_{C_4}(\lambda_1)$, we have

$$V_{C_4}(\lambda_1) \downarrow A_1 = 7, \quad V_{C_4}(\lambda_2) \downarrow A_1 = 12/8/4,$$
$$V_{C_4}(\lambda_3) \downarrow A_1 = 15/11/9/5/3, \text{ and } V_{C_4}(\lambda_4) \downarrow A_1 = 16/10/8/4.$$

Now let $V = 3 \otimes 3^{(q)}$, so $X < Y = C_2 \otimes C_2 < D_8$. If C is one of the factors C_2, then $C < A_7$ as before, and we see that

$$W_1 \downarrow C = 10^{16}/11^4, \quad W_2 \downarrow C = 20^5/01^{10}/02/00^{14}.$$

Hence

$$W_1 \downarrow C_2C_2 = 10 \otimes 11/11 \otimes 10, \quad W_2 \downarrow C_2C_2 = 20 \otimes 01/01 \otimes 20/02 \otimes 00/00 \otimes 02.$$

To find $W_i \downarrow X$ we must find the composition factors of the restrictions of $W_{C_2}(\lambda)$ to a regular A_1 in C_2, for $\lambda = 10, 01, 20, 02$ and 11. Easy calculations show that these restrictions are $3, 4, 6/2, 8/4$ and $7/5/1$, respectively.

Finally, consider $V = 1 \otimes 1^{(q)} \otimes 1^{(q')} \otimes 1^{(q'')}$, with $X < Y = C_1 \otimes C_1 \otimes C_1 \otimes C_1 < D_8$. Each of the factors C_1 lies in a Levi subgroup A_7, and as above,

$$W_1 \downarrow C_1 = 3^8/1^{48}, \quad W_2 \downarrow C_1 = 4/2^{27}/0^{42}.$$

The conclusion follows easily for this case. $\square$

Proposition 2.14 Let X be A_n $(n \le 4)$, B_n $(n \le 3)$, G_2 or D_4. The composition factors of various tensor products of Weyl modules for X are given as follows (as always, each factor given represents a Weyl module $W_X(\lambda)$):

$X = A_2\,(p \ne 2,3):$ $10 \otimes 10 = 20/01, 10 \otimes 01 = 11/00, 10 \otimes 02 = 12/01,$
$10 \otimes 20 = 30/11, 10 \otimes 11 = 21/02/10, 10 \otimes 21 = 31/12/20,$
$10 \otimes 12 = 22/03/11, 10 \otimes 22 = 32/13/21,$
$11 \otimes 11 = 22/30/03/11^2/00, 20 \otimes 02 = 22/11/00$

$X = B_2\,(p \ne 2,3):$ $01 \otimes 01 = 02/10/00, 10 \otimes 10 = 20/02/00, 10 \otimes 01 = 11/01,$
$01 \otimes 02 = 03/11/01, 10 \otimes 02 = 12/02/10,$
$01 \otimes 11 = 12/02/20/10$

$X = G_2\,(p \ne 2):$ $10 \otimes 10 = 20/01/10/00$

$X = A_3\,(p \ne 2):$ $100 \otimes 010 = 110/001, 010 \otimes 010 = 020/101/000$

$X = B_3\,(p \ne 2):$ $100 \otimes 100 = 200/010/000, 001 \otimes 001 = 002/010/100/000,$
$100 \otimes 001 = 101/001$

$X = A_4:$ $1000 \otimes 0100 = 1100/0010, 1000 \otimes 0010 = 1010/0001$

$X = D_4:$ $1000 \otimes 0010 = 1010/0001$

Proof. This is proved as in 2.10 by listing the weights of each factor in the tensor product (for which [BW] is useful), and considering sums of such weights. □

3 Complements in parabolics: proof of Theorem 1

In this section we establish Theorem 1. At the end of the section we also prove a version of Theorem 1 for classical groups of rank at most 8 (see Theorem 3.8).

Assume the hypotheses of Theorem 1. Thus G is a simple algebraic group of exceptional type over an algebraically closed field K of characteristic p, and X is a closed connected reductive subgroup of G. Moreover, $p = 0$ or $p > N(X, G)$, and X lies in a parabolic subgroup $P = QL$ of G with unipotent radical Q and Levi subgroup L.

Suppose now that X is simple and that P is a minimal parabolic subgroup containing X. The general result follows easily from this case, as we shall show at the end of the proof.

Lemma 3.1 *The L'-composition factors within Q have the structure of high weight modules for L'. If L_0 is a simple factor of L', then the possible high weights λ of nontrivial L_0-composition factors are as follows:*

(i) $L_0 = A_n$: $\lambda = 2\lambda_1, 2\lambda_n, 3\lambda_1, \lambda_j$ or λ_{n+1-j} $(j = 1, 2, 3)$ (note that $2\lambda_1, 2\lambda_n$ only occur if $G = F_4$ and $n \leq 2$, and $3\lambda_1$ only if $G = G_2$ and $n = 1$);

(ii) $L_0 = B_n$ or C_n $(n = 2$ or 3, $G = F_4)$: $\lambda = \lambda_1, \lambda_2$ or λ_3;

(iii) $L_0 = D_n$: $\lambda = \lambda_1, \lambda_{n-1}$ or λ_n;

(iv) $L_0 = E_6$: $\lambda = \lambda_1$ or λ_6;

(v) $L_0 = E_7$: $\lambda = \lambda_7$.

Proof. The high weight module structure is established in [ABS]. The list of possible high weights can be deduced from Theorem 2 of [ABS] (see also Remark 1 after [ABS, Theorem 3]). $\square$

The next two lemmas determine the embedding of X in L' modulo Q. For convenience, define $\bar{X}$ to be the simple closed connected subgroup such that

$$XQ = \bar{X}Q \text{ and } \bar{X} \leq L'$$

(so $\bar{X} \cong X$ as abstract groups).

Recall the definition of essential embedding given in the Introduction (just before the statement of Theorem 5). We make one further definition: if D is a classical simple algebraic group, we say that a semisimple closed connected subgroup X of D is *non-parabolically embedded* in D if X lies in no proper parabolic subgroup of D. Specifically, this means that if $V = V_D(\lambda_1)$ is the natural module, then one of the following holds:

$D = A_n$ and X is irreducible on V;

$D = B_n, C_n$ or D_n and $V \downarrow X = V_1 \perp \ldots \perp V_k$ with the V_i all non-degenerate, irreducible and inequivalent as X-modules;

$D = D_n$, $p = 2$ and $V \downarrow X = V_1 \perp \ldots \perp V_k$ with the V_i all non-degenerate, $V_2 \downarrow X, \ldots, V_k \downarrow X$ irreducible and inequivalent, and $V_1 \downarrow X$ giving a non-parabolic embedding of X in B_{m-1}, where dim $V_1 = 2m$.

Lemma 3.2 *The embedding of $\bar{X}$ in L' satisfies one of the following conditions:*

(i) *there is a maximal rank subgroup M of L' such that $\bar{X} \leq M$, M is a commuting product of classical simple groups, and the projection of $\bar{X}$ in each simple factor of M is non-parabolically embedded;*

(ii) *$L' = E_6$ or E_7, and either $\bar{X} = L'$ or $\bar{X}$ is a maximal connected subgroup of L' and is not a subsystem group;*

(iii) *$L' = E_7$, $\bar{X} = SL_2$ and $Z(L') \leq Z(\bar{X})$;*

(iv) *$L' = E_7$, $\bar{X} = A_1$ and $\bar{X}$ is essentially embedded in a maximal connected subgroup A_2 or G_2C_3 of L';*

(v) *$L' = E_6$ or A_1E_6, $\bar{X} = A_1$ and the projection of $\bar{X}$ to the factor E_6 is essentially embedded in a maximal subgroup C_4, F_4, A_2, G_2 or A_2G_2 of E_6, or in a subgroup A_1G_2 in $F_4 < E_6$.*

Proof. By the minimal choice of P, $\bar{X}$ lies in no proper parabolic subgroup of L'. Hence if L' is a product of classical groups, then conclusion (i) holds. Therefore we may assume that L' has an exceptional factor, so that L' is E_6, A_1E_6 or E_7.

If $C_{L'}(\bar{X}) \neq Z(L')$ then $\bar{X} \leq C_{L'}(t)$ for some semisimple element t of $L' - Z(L')$, and by the minimality of P, $C_{L'}(t)^0$ must be a product of classical groups, yielding (i). So we can assume that $C_{L'}(\bar{X}) = Z(L')$. We may also suppose that $\bar{X} \neq L'$ (otherwise (ii) holds), and choose a maximal connected subgroup M of L' containing X. Then M is non-parabolic, by the minimality of P.

Thus far, we have

$$L' = E_6, A_1E_6 \text{ or } E_7, \ \ C_{L'}(\bar{X}) = Z(L'), \ \ \bar{X} \leq M \text{ max } L'.$$

If M is a subsystem subgroup of L' then (i) holds (again by the minimality of P), so we assume that M is not a subsystem group. Thus M is given by [Se2, Theorem 1] (note that [Se2] applies for M, since $p > N(X, G)$ implies $p > N(M, G)$ except when $(X, M) = (A_2, G_2)$ and $G = E_7$ or E_8; but for this case, $p = 0$ or $p > 5$, so [Se2, Theorem 1] applies anyway). Notice that by our hypothesis that $p = 0$ or $p > N(X, G)$, we have $p \neq 2, 3, 5$ if $X = A_2$ and $p \neq 2, 3, 5, 7$ if $X = A_1$ or G_2.

First consider $L' = E_7$. Here $M = A_1$, A_2, A_1A_1, A_1G_2, A_1F_4 or G_2C_3. If $M = A_1A_1$, A_1G_2 or A_1F_4, then $\bar{X} = A_1$ and $\bar{X}$ projects nontrivially to both factors of M. Hence we see from the restrictions $V_{L'}(\lambda_7) \downarrow M$ given in 2.5 that $\bar{X} = SL_2$ and the involution in $Z(\bar{X})$ acts as -1 on $V_{L'}(\lambda_7)$, so (iii) holds. If $M = G_2C_3$ then since $G_2 \not\leq C_3$ (as $p \neq 2$), we have $\bar{X} = A_1$ or A_2. In the latter case $\bar{X} = SL_3$ (since any A_2 in G_2 is SL_3), so $C_{L'}(\bar{X}) \neq Z(L')$, a contradiction. And when $\bar{X} = A_1$, the condition $C_{L'}(\bar{X}) = Z(L')$ forces $\bar{X}$ to be essential in G_2C_3, so (iv) holds. Finally, when $M = A_1$ or A_2, conclusion (ii) or (iv) holds.

Now let $L' = E_6$ or A_1E_6. Then $M \cap E_6$ is A_2, G_2, F_4, C_4 or A_2G_2. Suppose $M \cap E_6 \neq F_4$. Then the condition $C_{L'}(\bar{X}) = Z(L')$, together with 2.8, forces either $\bar{X} = M$, or $\bar{X} = A_1$ with essential projection in $M \cap E_6$; so (v) holds. To conclude, suppose $M \cap E_6 = F_4$. If $\bar{X} = F_4$ then (ii) holds. And if $\bar{X} \neq F_4$, let M_0 be a maximal connected subgroup of $M \cap E_6$ containing the projection of $\bar{X}$. Then by [Se2, Theorem 1], M_0 is A_1 or A_1G_2, and (v) holds. $\square$

Using 3.2, together with 2.8, it is routine to list explicitly all the possible embeddings of $\bar{X}$ in L', and we do this for $\bar{X} \neq A_1$ in the next lemma.

Lemma 3.3 *Suppose that $\bar{X} \neq A_1$. Then the possibilities for L' and the embedding of $\bar{X}$ in L' are as in the following table. Further, the irreducible projections of $\bar{X}$ in the factors of L' are given by 2.8.*

$\bar{X}$	possibilities for L'	embedding of $\bar{X}$ in L'
A_2	A_2, A_2^2, A_5, D_4, A_2D_4, A_7	each projection irreducible
	E_6	$\bar{X} < A_2^3 < E_6$, or $\bar{X}$ max E_6
	E_7	$\bar{X} < A_2A_5$ or A_7, or $\bar{X}$ max E_7
B_2	B_2, A_3, A_4, A_3^2, A_3A_4, D_5, D_7	each projection irreducible
	D_5	$\bar{X} < B_2B_2 < D_5$
G_2	B_3, A_6, D_7	each projection irreducible
	D_4, D_7	$\bar{X} < B_3$, B_3B_3, resp.
	E_6	$\bar{X}$ max E_6
A_3	B_3, $C_3\,(p=2)$	orthogonal
	A_3, A_5, A_3^2	each projection irreducible
	D_6	$\bar{X} < D_3D_3$
B_3	B_3, D_4, A_6, A_7	$\bar{X}$ irreducible
	D_4	$\bar{X}$ reducible
	D_7	$\bar{X} < B_3B_3$
	E_7	$\bar{X} < A_7 < E_7$
C_3	C_3, A_5, D_7	$\bar{X}$ irreducible
$A_r\,(r \geq 4)$	A_r (or E_7, $r = 7$)	
$B_r\,(r \geq 4)$	D_{r+1}	natural

C_4	A_7	natural
	E_6	$\bar{X}$ max E_6
	E_7	$\bar{X} < A_7 < E_7$
D_4	D_4, A_7	natural
	E_7	$\bar{X} < A_7 < E_7$
$D_r\ (r \geq 5)$	D_r	
F_4	E_6	$\bar{X}$ max E_6
$E_r\ (r = 6, 7, 8)$	E_r	

Now let

$$1 = Q_0 < Q_1 < \ldots < Q_m = Q$$

be an L'-composition series for Q, with each factor $V_i = Q_i/Q_{i-1}$ an irreducible high weight module for L'.

The next two lemmas contain the heart of the proof of Theorem 1.

Lemma 3.4 *Suppose that $X \neq A_1$, and let V be an $\bar{X}$-composition factor of $V_i \downarrow \bar{X}$ for some i. Then one of the following holds:*

(i) in the semidirect product $\bar{X}V$, there is just one conjugacy class of closed complements to V;

(ii) $\bar{X}$, V and P are as follows:

$\bar{X}$	V	P
$B_r\ (r \geq 5)$	$V_{\bar{X}}(\lambda_1),\ p = 2$	D_{r+1}-*parabolic*
C_3	$V_{\bar{X}}(\lambda_2),\ p = 3$	A_5-*parabolic*
F_4	$V_{\bar{X}}(\lambda_4),\ p = 3$	E_6-*parabolic*

Proof. In the table below, we record the following information for each of the embeddings $\bar{X} < L'$ given in Lemma 3.3:

1. the possible nonzero high weights of the L'-composition factors V_i within Q (which are given by 3.1 when L' is simple, and using 2.1 and 2.3 if L' is not simple);

2. the possible nonzero high weights of the $\bar{X}$-composition factors of $V_i \downarrow \bar{X}$ (which are given by appropriate references in Section 2, listed in the last column of the table);

3. a lower bound for $N = N(X, G)$, as assumed in the hypothesis of Theorem 1.

$\bar{X}$	L'	$N \geq$	high wts. for $Q \downarrow L'$ (listed up to duals)	high wts. for $Q \downarrow \bar{X}$ (up to duals and twists)	reference
A_2	A_2	3	$\lambda_1, 2\lambda_1$	$10, 20$	
	A_2^2	3	$\lambda_i \otimes \lambda_j$	$10, 20, 11, 10 \otimes 10^{(q)},$ $10 \otimes 01^{(q)}$	2.14
	A_5	3	$\lambda_1, \lambda_2, \lambda_3$	$20, 30, 21$	2.10
	D_4	3	$\lambda_1, \lambda_3, \lambda_4$	11	
	$A_2 D_4$	5	$\lambda_i \otimes 0, 0 \otimes \lambda_j,$ $\lambda_i \otimes \lambda_j$ $(j \neq 2)$	$10, 20, 11, 21,$ $10 \otimes 11^{(q)}, 11 \otimes 10^{(q)}$	2.14
	A_7	5	$\lambda_1, \lambda_2, \lambda_3$	$11, 30, 22$	2.10
	E_6	5	λ_1	$10, 20, 11, 10 \otimes 10^{(q)},$ $10 \otimes 01^{(q)}$ if $\bar{X} < A_2^3;$ 22 if $\bar{X}$ max L'	2.3 2.5
	E_7	5	λ_7	$10, 11, 21, 30, 10 \otimes 20^{(q)},$ $20 \otimes 10^{(q)}, 10 \otimes 02^{(q)},$ $02 \otimes 10^{(q)},$ if $\bar{X} < A_2 A_5;$ $30, 11$ if $\bar{X} < A_7;$ 60 if $\bar{X}$ max L'	2.3, 2, 14 2.3, 2.10 2.5
B_2	B_2	2	$\lambda_1, \lambda_2, 2\lambda_2$	$10, 01, 02$	
	A_3	3	λ_1, λ_2	$10, 01$	
	A_4	3	λ_1, λ_2	$10, 02$	2.10
	A_3^2	5	$\lambda_i \otimes 0, 0 \otimes \lambda_i,$ $\lambda_i \otimes \lambda_j$ $((i,j) \neq (2,2))$	$10, 01, 02, 11, 10 \otimes 01^{(q)},$ $01 \otimes 10^{(q)}, 01 \otimes 01^{(q)}$	2.14
	$A_3 A_4$	5	$\lambda_i \otimes 0, 0 \otimes \lambda_j,$ $\lambda_1 \otimes \lambda_j, \lambda_i \otimes \lambda_1$	$10, 01, 20, 02, 03, 11,$ $10 \otimes 10^{(q)}, 10 \otimes 01^{(q)},$ $01 \otimes 10^{(q)}, 01 \otimes 02^{(q)},$ $02 \otimes 01^{(q)}$	2.14
	D_7	5	$\lambda_1, \lambda_6, \lambda_7$	$20, 13, 03 (p = 7)$	2.12
	D_5	3	$\lambda_1, \lambda_4, \lambda_5$	$01, 02, 11$ if $\bar{X}$ irred. $10, 01, 02,$ $01 \otimes 01^{(q)}$ if $\bar{X} < B_2 B_2$	2.12 2.7, 2.14
G_2	B_3	2	λ_1, λ_3	10	2.12
	A_6	7	$\lambda_1, \lambda_2, \lambda_3$	$10, 20, 01$	2.10
	D_4	3	$\lambda_1, \lambda_3, \lambda_4$	10	
	D_7	7	$\lambda_1, \lambda_6, \lambda_7$	$01, 11$ if $\bar{X}$ irred. $10 \otimes 10^{(q)}$ if $\bar{X} < B_3 B_3$	2.12 2.7, 2.14
	E_6	7	λ_1	20	2.5
A_3	B_3, C_3	—	λ_1, λ_3	$100, 010$	
	A_3	—	λ_1, λ_2	$100, 010$	
	A_5	2	$\lambda_1, \lambda_2, \lambda_3$	$010, 101, 200$	2.10

	A_3^2	2	$\lambda_i \otimes 0, 0 \otimes \lambda_i$ $\lambda_i \otimes \lambda_j$ $((i,j) \neq (2,2))$	$100, 010, 200, 101, 110,$ $100 \otimes 100^{(q)}, 100 \otimes 001^{(q)},$ $100 \otimes 010^{(q)}, 010 \otimes 100^{(q)}$	2.14
	D_6	2	$\lambda_1, \lambda_5, \lambda_6$	$010, 100 \otimes 100^{(q)},$ $100 \otimes 001^{(q)}$	2.7
B_3	B_3	2	λ_1, λ_3	$100, 001$	
	D_4	2	$\lambda_1, \lambda_3, \lambda_4$	$100, 001$	
	A_6	2	$\lambda_1, \lambda_2, \lambda_3$	$100, 010, 002$	2.10
	A_7	2	$\lambda_1, \lambda_2, \lambda_3$	$100, 010, 001, 101$	2.10
	D_7	2	$\lambda_1, \lambda_6, \lambda_7$	$100, 001 \otimes 001^{(q)}$	2.7, 2.14
	E_7	2	λ_7	$100, 010$	2.3, 2.10
C_3	C_3	2	λ_3	001	
	A_5	2	$\lambda_1, \lambda_2, \lambda_3$	$100, 010, 001$	2.10
	D_7	3	$\lambda_1, \lambda_6, \lambda_7$	$010, 110, 100 (p=7)$	2.12
A_r	A_r	—	$\lambda_1, \lambda_2, \lambda_3$	$\lambda_1, \lambda_2, \lambda_3$	
B_4	D_5	2	$\lambda_1, \lambda_4, \lambda_5$	λ_1, λ_4	
B_r	D_{r+1}	—	$\lambda_1, \lambda_r, \lambda_{r+1}$	λ_1, λ_r	
C_4	A_7	2	$\lambda_1, \lambda_2, \lambda_3$	$1000, 0100, 0010$	2.10
	E_6	2	λ_1	0100	2.5
	E_7	2	λ_7	0100	2.3, 2.10
D_4	D_4	2	$\lambda_1, \lambda_3, \lambda_4$	$1000, 0010, 0001$	
	A_7	2	$\lambda_1, \lambda_2, \lambda_3$	$1000, 0100, 0011$	2.10
	E_7	2	λ_7	0100	2.3, 2.10
D_r	D_r	—	$\lambda_1, \lambda_{r-1}, \lambda_r$	$\lambda_1, \lambda_{r-1}, \lambda_r$	
F_4	E_6	—	λ_1	0001	2.5
E_6	E_6	—	λ_6	λ_6	
E_7	E_7	—	λ_7	λ_7	

Let $V = V_{\bar{X}}(\lambda)$ be an $\bar{X}$-composition factor occurring in $Q \downarrow \bar{X}$, so that λ is a twist of one of the weights listed in the fifth column of the table (up to duals). By 1.3 and 1.5, we may assume that no twist is involved (i.e. λ is in the table, up to duals).

Suppose first that λ is not restricted. Then V or V^* is one of the tensor products involving a twist in the table, and we deduce from 1.14 and 1.15 that conclusion (i) of the lemma holds.

Now suppose that λ is restricted and that λ is not as in conclusion (ii) of the lemma. Using 1.9 for $\bar{X} = A_2$, 1.10 and 1.14 for $\bar{X} = B_2$, and 1.11 and 1.12 for the remaining cases, we see that $W_{\bar{X}}(\lambda)$ has no trivial composition factors. Therefore conclusion (i) holds, by 1.7. □

Lemma 3.5 *Suppose that* $\bar{X} = A_1$, *and let* V *be a composition factor of* $V_i \downarrow \bar{X}$ *for some* i. *Then either conclusion (i) of Lemma 3.4 holds, or* $V = (p-2) \otimes 1^{(p)}$ *with* $p = 11$ *or* 17, *and* P *is a* D_7-*parabolic subgroup of* $G = E_8$.

Proof. Assume that (i) of 3.4 does not hold. Then by 1.7 and 1.8, V is a Frobenius twist of the $\bar{X}$-module $(p-2) \otimes 1^{(p)}$. Write $V = V_{\bar{X}}(\lambda)$.

The embedding of $\bar{X}$ in L' is given by 3.2. We divide the proof into four cases:

1. L' has a factor A_m with $m \geq 4$;

2. L' has a factor D_m with $m \geq 4$;

3. every factor of L' has rank 3 or less;

4. $L' = E_7$, E_6 or $A_1 E_6$.

Let V_i be an L'-composition factor in $Q \downarrow L'$ such that $V = V_{\bar{X}}(\lambda)$ occurs in $V_i \downarrow \bar{X}$. Then V_i is a tensor product of irreducible modules for the simple factors of L', which have high weights given by 3.1.

Case 1.   Here L' is A_m $(4 \leq m \leq 7)$, $A_6 A_1$, $A_5 A_1$, $A_4 A_r$ $(r \leq 3)$, $A_4 A_2 A_1$ or $A_4 A_1^2$, and V_i is a tensor product of modules for the factors with high weights 0, λ_1, λ_2 or λ_3 (or duals of these). By 2.9, the projection of $\bar{X}$ in the factor A_m $(m \geq 4)$ gives a representation of $\bar{X}$ of high weight μ, where (up to twists) either $\mu = m$, or $m = 5$ and $\mu = 2 \otimes 1^{(q)}$ or $1 \otimes 2^{(q)}$, or $m = 7$ and $\mu = 1 \otimes 3^{(q)}$, $3 \otimes 1^{(q)}$ or $1 \otimes 1^{(q)} \otimes 1^{(q')}$. The composition factors of $V_{A_m}(\lambda_j) \downarrow \bar{X}$ $(j = 1, 2, 3)$ are given in 2.13.

Suppose now that $G = E_6$, so $p > 5$ by hypothesis. Then L' is A_5, A_4 or $A_4 A_1$. If $L' = A_5$ then $\mu = 5$, $2 \otimes 1^{(q)}$ or $1 \otimes 2^{(q)}$ and 2.13 implies that $V_{\bar{X}}(\lambda)$ does not occur in $\bigwedge^j V_{\bar{X}}(\mu)$ for $j = 1, 2, 3$, a contradiction. If $L' = A_4 A_1$ then $\mu = 4$, so $V_{\bar{X}}(\lambda)$ must occur in $\bigwedge^j V_{\bar{X}}(4) \otimes 1$ or $\bigwedge^j V_{\bar{X}}(4) \otimes 1^{(q)}$ $(j \leq 2)$. As $\bigwedge^2 V_{\bar{X}}(4) = 6/2$, this is not possible. Similarly $L' \neq A_4$.

Now assume that $G = E_7$ or E_8, so $p > 7$. If $m = 7$ (so $L' = A_7$) then μ is 7, $3 \otimes 1^{(q)}$, $1 \otimes 3^{(q)}$ or $1 \otimes 1^{(q)} \otimes 1^{(q')}$ and we see from 2.13 that $V_{\bar{X}}(\lambda)$ does not occur in $\bigwedge^j V_{\bar{X}}(\mu)$ for $j \leq 3$; and if $m = 4$ or 6 then $\mu = m$ and again 2.13 implies that $V_{\bar{X}}(\lambda)$ does not occur in $\bigwedge^j V_{\bar{X}}(m) \otimes 1^{(q_1)} \otimes 1^{(q_2)}$, $\bigwedge^j V_{\bar{X}}(4) \otimes 2^{(q_1)} \otimes 1^{(q_2)}$ or $\bigwedge^j V_{\bar{X}}(4) \otimes r^{(q_1)}$ (where $q_1, q_2 \geq 1$ and $r \leq 3$). Hence $m = 5$, so that $L' = A_5$ or $A_5 A_1$ and μ is 5, $2 \otimes 1^{(q)}$ or $1 \otimes 2^{(q)}$. By 2.13, $V_{\bar{X}}(\lambda)$ does not occur in $\bigwedge^j V_{\bar{X}}(\mu)$ or $\bigwedge^j V_{\bar{X}}(\mu) \otimes 1$ for $j \leq 3$, and can only occur in $\bigwedge^j V_{\bar{X}}(\mu) \otimes 1^{(p)}$ if $\mu = 5$, $p = 11$ and $j = 3$. Therefore $L' = A_5 A_1$ and $Q \downarrow L'$ has a composition factor $V_{A_5}(\lambda_3) \otimes V_{A_1}(1)$. But [ABS] implies that there is no such composition factor in $Q \downarrow L'$.

Case 2.   In this case, L' has a factor D_m with $m \geq 4$, so L' is D_m, $D_5 A_r$ or $D_4 A_r$ $(r \leq 2)$. By the definition of non-parabolic embedding, the projection of $\bar{X}$

in the factor D_m gives an embedding of $\bar{X}$ in a subgroup M of D_m, where M is a product of orthogonal groups and $\bar{X}$ maps to an irreducible subgroup of each of these. Explicitly, the possibilities are as follows:

$$m = 7: \quad \bar{X} \hookrightarrow M = \quad B_6, \; B_5B_1, \; B_4B_2, \; B_3B_3, \; B_3B_1^2, \; B_2^2B_1, \; B_2B_1^3,$$
$$B_4D_2, \; B_3B_1D_2, \; B_1^3D_2, \; B_2D_4 \text{ or } B_1^2D_4$$

$$m = 6: \quad \bar{X} \hookrightarrow M = \quad D_6, \; B_5, \; B_4B_1, \; B_3B_2, \; B_2B_1^2, \; B_1^4, \; B_3D_2, \; , B_1D_2^2,$$
$$D_2^3, \; D_4D_2 \text{ or } B_1D_4$$

$$m = 5: \quad \bar{X} \hookrightarrow M = \quad B_4, \; B_3B_1, \; B_2B_2, \; B_1^3, \; B_2D_2 \text{ or } B_1^2D_2$$

$$m = 4: \quad \bar{X} \hookrightarrow M = \quad D_4, \; B_3, \; B_2B_1, \; B_1D_2 \text{ or } D_2D_2$$

Recall that $V = V_{\bar{X}}(\lambda)$ occurs in the restriction to $\bar{X}$ of the L'-composition factor V_i within Q. If V_i is a spin module for D_m, then by 2.5, the composition factors of $V_i \downarrow M$ are tensor products of spin modules for the factors of M.

Suppose first that $G = E_6$, so $p > 5$. Then $L' = D_4$ or D_5 and we see easily using 2.13 that $V_{\bar{X}}(\lambda)$ does not occur in $Q \downarrow \bar{X}$; for example, assume $L' = D_5$ and $\bar{X} \hookrightarrow M = B_3B_1$, so that $V_{L'}(\lambda_1) \downarrow \bar{X} = 6^{(q_1)}/2^{(q_2)}$ (where $q_1, q_2 \geq 1$). By 3.1, V_i is $V_{L'}(\lambda_j)$ with $j = 1, 4$ or 5, and so by 2.13, $V_i \downarrow \bar{X}$ is $6^{(q_1)}/2^{(q_2)}$ or $(6^{(q_1)}/0) \otimes 1^{(q_2)}$. Hence $V_{\bar{X}}(\lambda)$ does not occur does not occur in $V_i \downarrow \bar{X}$.

Now let $G = E_7$ or E_8, so $p > 7$. If $m = 4$ or 5 then $L' = D_mA_r$ with $0 \leq r \leq 2$, and $V_i = V_{D_m}(\lambda_j) \otimes V_{A_r}(\lambda_k)$ with $j = 1, m$ or $m-1$. Again, 2.13 implies that $V_{\bar{X}}(\lambda)$ does not occur in $V_i \downarrow \bar{X}$; for example, suppose $m = 5$, $j = 4$ or 5 and $\bar{X} \hookrightarrow B_3B_1A_r$. Then $V_i \downarrow \bar{X} = (6/0)^{(q_1)} \otimes 1^{(q_2)} \otimes r^{(q_3)}$ for some powers q_1, q_2, q_3 of p (possibly equal to 1), and clearly $V_{\bar{X}}(\lambda)$ does not occur.

Thus $m = 6$ or 7, so $L' = D_m$. When $m = 6$, or $m = 7$ and $M \neq B_5B_1$, we check using 2.13 again that $V_{\bar{X}}(\lambda)$ does not occur in $V_i \downarrow \bar{X} = V_{D_m}(\lambda_j) \downarrow \bar{X}$ $(j = 1, m-1$ or $m)$. So $m = 7$ and $\bar{X} \hookrightarrow B_5B_1$. If V_i is a spin module for D_7, then by 2.13, $V_i \downarrow \bar{X} = (15/9/5) \otimes 1^{(q_1)}$ or $(15/9/5)^{(q_2)} \otimes 1$ (where $q_1, q_2 \geq 1$). The fact that $V_{\bar{X}}(\lambda)$, a twist of $(p-2) \otimes 1^{(p)}$, occurs in $V_i \downarrow \bar{X}$ forces the first possibility to hold, with $q_1 = p$ and $p = 11$ or 17, as in the conclusion of the lemma.

Case 3.   Here L' is a product of simple groups of rank 3 or less.

Suppose $G = F_4$, so $p > 3$. If $L' = B_3$ then $V_{L'}(\lambda_1) \downarrow \bar{X}$ is 6 or $1^{(q_1)} \otimes 1^{(q_2)}/2^{(q_3)}$ $(q_1, q_2, q_3 \geq 1)$, and $V_i = V_{L'}(\lambda_j)$ with $j = 1$ or 3; and if $L' = C_3$ then $V_{L'}(\lambda_1)$ is $5, 3^{(q_1)}/1^{(q_2)}, 2 \otimes 1^{(q)}, 1 \otimes 2^{(q)}$ or $1/1^{(q_1)}/1^{(q_2)}$, and $V_i = V_{L'}(\lambda_3)$, which lies in $\bigwedge^3 V_{L'}(\lambda_1)$. Hence in both cases, 2.13 shows that $V_{\bar{X}}(\lambda)$ does not occur in $V_i \downarrow \bar{X}$. The same conclusion is immediate for the other possibilities for L'.

In the remaining cases $G = E_6\,(p > 5)$ or $E_7, E_8\,(p > 7)$, L' is a product of factors A_n with $n \leq 3$. The projection of $\bar{X}$ in a factor A_n gives a twist of the representation with high weight n, or possibly of $1 \otimes 1^{(q)}$ with $n = 3$. Hence by 2.13, any composition factor of $V_i \downarrow \bar{X}$ is a tensor product of $\bar{X}$-modules, one for each factor A_n of L', which are twists of restricted modules of high weights at most

$4\,(n = 3)$, $2\,(n = 2)$ and $1\,(n = 1)$. It follows that $V_{\bar{X}}(\lambda)$ does not occur in $V_i \downarrow \bar{X}$.

Case 4.   In this last case, L' is E_7, E_6 or $E_6 A_1$, and $p > 7$. The embedding of $\bar{X}$ in L' is given by 3.2.

Suppose that 3.2(i) holds, so that $\bar{X}$ lies in a maximal rank subgroup M of L' which is a product of classical groups. If $L' = E_7$ then M is A_7, $A_1 D_6$, $A_1^3 D_4$ or $A_2 A_5$; and if $L' = E_6 A_1^r$ ($r = 0$ or 1), then M is $A_5 A_1^{r+1}$ or $A_2^3 A_1^r$. The nontrivial composition factor V_i of $Q \downarrow L'$ has high weight λ_7 if $L' = E_7$, λ_1 or λ_6 if $L' = E_6$, and $\lambda_1 \otimes 1$ or $\lambda_6 \otimes 1$ if $L' = E_6 A_1$. The composition factors of $V_i \downarrow M$ are given by 2.3. As before, $\bar{X}$ projects to an irreducible subgroup of any factor A_n of M; and if $M = A_1 D_6$ then $\bar{X}$ projects to one of the subgroups of D_6 listed in case 2 above.

Let W be a nontrivial composition factor of $V_i \downarrow M$. If $M = A_1 D_6$ then W is $1 \otimes \lambda_1$ or $0 \otimes \lambda_5$, and we see as in case 2 that $V_{\bar{X}}(\lambda)$ does not occur in $W \downarrow \bar{X}$. If $M = A_7$ then $W = V_M(\lambda_j)$ with $j = 2$ or 6, so $W \downarrow \bar{X}$ is $\bigwedge^2 V_{\bar{X}}(7)$, $\bigwedge^2 V_{\bar{X}}(3 \otimes 1^{(q)})$ or $\bigwedge^2 V_{\bar{X}}(1 \otimes 1^{(q)} \otimes 1^{(q')})$, and by 2.13, $V_{\bar{X}}(\lambda)$ does not occur in any of these modules. Similarly, 2.13 rules out the other possibilities for M.

Now consider 3.2(ii). Here $\bar{X}$ is a maximal connected non-subsystem subgroup of L', so $L' = E_7$ by [Se2, Theorem 1]. By 2.5, $V_{\bar{X}}(\lambda)$ does not occur in $V_i \downarrow \bar{X}$. And in case (iii) of 3.2, conclusion (i) of 3.4 holds, by 1.13.

Next suppose that 3.2(iv) holds, so that $L' = E_7$ and $\bar{X}$ is essential in a maximal connected subgroup $M = A_2$ or $G_2 C_3$ of L'. Let W be a nontrivial composition factor of the restriction $V_i \downarrow M$, which is given by 2.5. If $M = G_2 C_3$ then W is $10 \otimes 100$ or $00 \otimes 001$; in the first case, $W \downarrow \bar{X} = 6^{(q_1)} \otimes 5^{(q_2)}$, and in the second, $W \downarrow \bar{X} = 9^{(q_1)}/3^{(q_2)}$, so $V_{\bar{X}}(\lambda)$ does not occur. If $M = A_2$ then $W = 60$ or 06, which is the sixth symmetric power of 10 or 01. Hence the highest weight of $\bar{X}$ occurring in $W \downarrow \bar{X}$ is 12, and so $V_{\bar{X}}(\lambda)$ does not occur.

Finally, consider case (v) of 3.2. Here L' is E_6 or $E_6 A_1$, and $\bar{X}$ is essential in M or $M A_1$, where M is a subgroup C_4, F_4, A_2, G_2, $A_2 G_2$ or $A_1 G_2$ of E_6. The restriction $V_{E_6}(\lambda_1) \downarrow M$ is given by 2.5 except for the last subgroup $M = A_1 G_2$ (and in this case, $M < A_2 G_2$ so $V_{E_6}(\lambda_1) \downarrow M$ can be deduced from $V_{E_6}(\lambda_1) \downarrow A_2 G_2$). When M is simple, 2.5 and 2.13 show that $V_{\bar{X}}(\lambda)$ does not occur in $V_i \downarrow \bar{X}$. Now consider $M = A_2 G_2$ and let π be the projection of $\bar{X}$ to M. Here, $V_{E_6}(\lambda_1) \downarrow \bar{X}\pi$ is $2^{(q_1)} \otimes 6^{(q_2)}/4^{(q_1)} \otimes 0/0$ ($q_1, q_2 \geq 1$), and it follows that $V_{\bar{X}}(\lambda)$ does not occur in $V_i \downarrow \bar{X}$. Similarly, this conclusion holds when $M = A_1 G_2$.

This completes the proof of Lemma 3.5. □

To finish the proof of Theorem 1(under our initial assumptions that X is simple and P is a minimal parabolic containing X), it remains to handle the exceptional situations which occur in 3.4 and 3.5.

Lemma 3.6 *Theorem 1 (with X simple, P minimal) holds if $X \neq A_1$.*

Proof. If conclusion (i) of 3.4 holds for all composition factors V of $Q \downarrow \bar{X}$, then

$Q\bar{X}$ has just one Q-conjugacy class of closed complements to Q, as required for Theorem 1.

Thus we suppose that (ii) of 3.4 holds for some composition factor V of $Q \downarrow \bar{X}$. Here $\bar{X}$ is B_r ($r \geq 5$), C_3 or F_4 and L' is D_{r+1}, A_5 or E_6, respectively. Moreover, if V_i is the composition factor of $Q \downarrow L'$ such that $V_i \downarrow \bar{X}$ has composition factor V as in 3.4(ii), then V_i is $V_{D_{r+1}}(\lambda_1)$, $V_{A_5}(\lambda_2)$ or $V_{E_6}(\lambda_j)$ ($j = 1$ or 6), respectively. Then by 1.16, there is just one conjugacy class of closed complements to V_i in the semidirect product $\bar{X}V_i$, and so Theorem 1 holds again. $\square$

Lemma 3.7 *Theorem 1 (with X simple, P minimal) holds if $X = A_1$.*

Proof. As in the previous proof, we can suppose that 3.4(i) does not hold for some composition factor V of $Q \downarrow \bar{X}$. Then by 3.5, we have $G = E_8$, $L' = D_7$, $V = (p-2) \otimes 1^{(p)}$ and $p = 11$ or 17. Moreover, from the proof of 3.5 (Case 2), we see that $\bar{X} < B_5 B_1 < L'$, with $V_{L'}(\lambda_1) \downarrow \bar{X} = 10/2^{(p)}$, and

$$V_{L'}(\lambda_6) \downarrow \bar{X} = 15 \otimes 1^{(p)}/9 \otimes 1^{(p)}/5 \otimes 1^{(p)}, \text{ if } p = 17$$
$$V_{L'}(\lambda_6) \downarrow \bar{X} = 4 \otimes 1^{(p)} \otimes 1^{(p)}/5 \otimes 1^{(p)}/9 \otimes 1^{(p)}/5 \otimes 1^{(p)}, \text{ if } p = 11.$$

Now $L(G) = L(Q) + L(Q^-) + L(D_7 T_1)$ (where Q^- denotes the unipotent radical of the parabolic opposite to P). We calculate $L(G) \downarrow \bar{X}$.

We have $L(Q) \downarrow L' = \lambda_6/\lambda_1$, so $L(Q) \downarrow \bar{X}$ and $L(Q^-) \downarrow \bar{X}$ can be obtained from $V_{L'}(\lambda_6) \downarrow \bar{X}$ above. Also,

$$L(D_7) \downarrow \bar{X} = L(B_5) \downarrow \bar{X} \oplus L(B_1) \downarrow \bar{X} \oplus 10 \otimes 2^{(p)}$$
$$= \wedge^2 V_{\bar{X}}(10) \oplus \wedge^2 V_{\bar{X}}(2^{(p)}) \oplus 10 \otimes 2^{(p)}.$$

Hence, using 2.13, we deduce that

$$L(G) \downarrow \bar{X} = (15 \otimes 1^{(p)})^2/(9 \otimes 1^{(p)})^2/(5 \otimes 1^{(p)})^2/10 \otimes 2^{(p)}/1 \otimes 1^{(p)}/$$
$$14^2/10^3/6/2/(2^{(p)})^3/0, \text{ if } p = 17$$
$$L(G) \downarrow \bar{X} = (9 \otimes 1^{(p)})^2/(5 \otimes 1^{(p)})^4/(4 \otimes 2^{(p)})^2/7 \otimes 1^{(p)}/3 \otimes 1^{(p)}/10 \otimes 2^{(p)}/$$
$$10^3/6^2/4^2/2^2/(2^{(p)})^3/0, \text{ if } p = 11.$$

We now revert to the subgroup X of QL' given in the hypothesis of Theorem 1 (recall that $\bar{X}$ was chosen such that $\bar{X} \leq L'$ and $\bar{X}Q = XQ$). The restriction $L(G) \downarrow X$ has the same composition factors as $L(G) \downarrow \bar{X}$.

We have $X < QB_5 B_1$, where $B_5 B_1 < L'$. Moreover, the derived group Q' is L'-isomorphic to the usual module $V_{L'}(\lambda_1)$, so

$$Q' \downarrow B_5 B_1 = Q_{11} \times Q_3$$

where Q_{11}, Q_3 are the usual modules for B_5, B_1, and $Q_{11} \downarrow X = 10$, $Q_3 \downarrow X = 2^{(p)}$.

Define $A = C_{L(G)}(L(X))$. We claim that

$$(*) \quad \dim A \geq 4 \text{ and } A \cap L(Q) = L(Q_3).$$

To see this, observe that from the restrictions given above, the only composition factors in $C_{L(G)}(L(X))$ have high weights 0 or $2^{(p)}$. The only composition factor of $L(G) \downarrow X$ which extends $V_X(2^{(p)})$ or $V_X(0)$ indecomposably is $(p-2) \otimes 1^{(p)}$ (see [AJL, 3.9]). We now argue that $L(G) \downarrow X$ has a submodule which is X-isomorphic to either $2^{(p)} \oplus 0$ or $2^{(p)} \oplus 2^{(p)}$. Certainly $L(Q_3) \downarrow X = 2^{(p)}$, so we can suppose there are no further X-submodules $2^{(p)}$ or 0. Now $L(G)$ contains three linearly independent vectors having weight $2p$ for X, and each is a maximal vector for X, hence generates (under X) an image of the Weyl module $W_X(2p) = 0/(p-2) \otimes 1^{(p)}/2^{(p)}$. By assumption, none of these images has a trivial submodule, and two of them involve $(p-2) \otimes 1^{(p)}$. As $L(G) \downarrow X$ has just two composition factors $(p-2) \otimes 1^{(p)}$, this means that $L(G) \downarrow X$ has a trivial 1-dimensional quotient. Since $L(G)$ is self-dual, it follows that X fixes a 1-space, contrary to assumption.

This establishes our claim that $L(G) \downarrow X$ has a submodule which is X-isomorphic to either $2^{(p)} \oplus 0$ or $2^{(p)} \oplus 2^{(p)}$. As $L(X)$ acts trivially on this submodule, the submodule is contained in A, and so $\dim A \geq 4$. Finally, $L(Q) \downarrow X$ has just one composition factor $2^{(p)}$ and no trivial factors, so $C_{L(Q)}(L(X))$ has dimension 3 and is equal to $L(Q_3)$, proving $(*)$.

Next, define $R = C_G(A)$, so that $X \leq N_G(R)$. We analyse the group R. Let T be a maximal torus of X. From $L(G) \downarrow X$ we see that the number of weights of $L(G) \downarrow T$ which are divisible by p is 20 if $p = 17$, and 28 if $p = 11$. Therefore $C_{L(G)}(L(T))$ has dimension 20 $(p = 17)$ or 28 $(p = 11)$. Moreover, $C_{L(G)}(L(T)) = L(D)$ for some maximal rank reductive subgroup D of G. Since D has dimension 20 or 28, we must have

$$D = T_2 A_1^6, T_3 A_2 A_1^3, T_4 A_2^2 \text{ or } T_5 A_3, \text{ if } p = 17$$
$$D = T_2 A_1 A_2 A_3, T_4 A_4 \text{ or } T_1 A_1 A_2^3, \text{ if } p = 11$$

(note that $A_2^2 A_1^4$ is not contained in G). As $A \subseteq L(D)$ we have $Z(D) \leq C_G(A) = R$. Consequently R contains either T_2 (a rank 2 torus) or $T_1 \times Z_3$ (if $D = T_1 A_1 A_2^3$).

We now claim that $R \leq QB_5$. To see this, write $C = C_G(L(Q_3))$, and observe that C contains QB_5 and $R \leq C$. Now X normalizes C, and there is no proper reductive subgroup of G containing QB_5, so the subgroup CX is not reductive. Hence CX lies in a proper parabolic subgroup $\tilde{P}$ of G. As $QB_5 X = QB_5 B_1$ lies in $\tilde{P} \cap P$, we must have $P = \tilde{P}$, and so $C \leq P$. This gives $C = C_P(L(Q_3)) = QB_5$, and so $R \leq C = QB_5$, as claimed.

We have now established that

$$R \leq QB_5, \ X \leq N_P(R) \text{ and } R \text{ contains } T_2 \text{ or } T_1 \times Z_3.$$

It can be seen easily, using 2.8, that a regular subgroup A_1 in B_5 is maximal, so there is no proper $\bar{X}$-invariant subgroup of B_5 which contains T_2 or $T_1 \times Z_3$. Hence

$R/R \cap Q \cong B_5$. Let $Q_0 = R \cap Q$, so Q_0 is a B_5-invariant subgroup of Q. As B_5 is irreducible on $Q/Q' \cong V_{B_5}(\lambda_5)$, either $Q_0 = Q$ or $Q_0 \leq Q'$.

Suppose $Q_0 = Q$. Then $Q < R$, so $A \leq C_{L(G)}(Q)$. Now $C_{L(G)}(Q)$ is P-invariant, from which we see that $C_{L(G)}(Q) = L(Q')$. But this implies that $A \subseteq L(Q')$, contrary to $(*)$.

Thus $Q_0 \leq Q'$. We have $RQ'/Q' \cong B_5$. By 1.7, there is just one class of closed complements to Q/Q' in QB_5/Q', so replacing R (and X) by a suitable Q-conjugate, we can take $Q'R = Q'B_5$. Then $X \leq N_P(Q'R) \leq Q'L$. The restriction $Q' \downarrow X$ has no composition factor $(p-2) \otimes 1^{(p)}$, so we deduce as before that there is just one Q'-class of closed complements to Q' in $Q'X$, and hence Theorem 1 holds. □

Lemmas 3.6 and 3.7 complete the proof of Theorem 1 when X is simple and P is a minimal parabolic subgroup containing X.

Suppose now that the hypotheses of Theorem 1 hold, with X simple, and P *not* a minimal parabolic, so that $X < P_1 < P$ for some minimal parabolic P_1 containing X. We may take $P_1 = Q_1 L_1$, $P = QL$ with $Q < Q_1$, $L_1 < L$. Let $\bar{X}$ be the simple closed connected subgroup of L_1 such that $Q_1 X = Q_1 \bar{X}$.

In 3.4 and 3.5, it is proved that if V is a composition factor occurring in $Q_1 \downarrow \bar{X}$, then either

(i) there is just one class of closed complements to V in $\bar{X}V$, or

(ii) $\bar{X}$, V are as in 3.4(ii)

(note that $L_1' \neq D_7$ since $L_1' < L'$, so the exceptional case in 3.5 does not occur). If (i) holds for all V, then it certainly holds for all composition factors of $Q \downarrow \bar{X}$, so the conclusion of Theorem 1 follows. And in case (ii), the argument of 3.6 gives Theorem 1.

To complete the proof of Theorem 1, we must relax the assumption that X is simple. Suppose first that X is semisimple, say $X = X_1 \ldots X_t$ with each X_i simple and $t \geq 2$. By hypothesis, either $p = 0$ or $p > N(X, G) = \max(N(X_i, G) : 1 \leq i \leq t)$. Assume that X lies in a parabolic subgroup $P = QL$ of G.

Suppose Theorem 1 does not hold, so that the semidirect product XQ has more than one Q-conjugacy class of closed complements to Q. Then there is a composition factor W in $Q \downarrow X$ such that XW has more than one class of closed complements . Write $W = W_1 \otimes \ldots \otimes W_t$, where each W_i is an irreducible X_i-module, and assume that W_1 is not the trivial X_1-module. Then by 1.5, W_1 has a rational indecomposable extension by the trivial X_1-module (note that $(X_i, p) \neq (C_n, 2)$ for all i, by our hypothesis on p). By the proofs of 3.4 and 3.5, we conclude that the possibilities for X_1, W_1 and a minimal parabolic $Q_0 L_0$ of P containing X_1 are as follows:

(i) $(X_1, p) = (C_3, 3)$, $W_1 = V_{X_1}(\lambda_2)$, $L_0' = A_5$;

(ii) $(X_1, p) = (F_4, 3)$, $W_1 = V_{X_1}(\lambda_4)$, $L_0' = E_6$;

(iii) $(X_1, p) = (A_1, 11 \text{ or } 17)$, $W_1 = (p-2) \otimes 1^{(p)}$, $L_0' = D_7$.

In case (i), $G = E_7$ (not E_6, as $V_{A_5}(\lambda_2)$ occurs in Q); and in case (ii), $G = E_7$ or E_8. As $p = 3$, we have rank$(X_2) \geq 3$. But then in either case, $X_1 X_2$ cannot lie in a parabolic subgroup of G. In case (iii), $L' = L_0' = D_7$; but then $C_{L'Q/Q}(X_1Q/Q)^0 = 1$, contradicting the fact that X_2Q/Q lies in this centralizer.

Theorem 1 is now proved for X semisimple. Suppose finally that X is reductive, so that $X = X'T$ where $T = Z(X)^0$. Assume that X lies in a parabolic subgroup $P = QL$ of G. Let $Y = Y'S$ be a closed complement to Q in XQ (where $S = Z(Y)^0$). By Theorem 1 for the semisimple case, $Y' = (X')^h$ for some $h \in Q$. Hence $S \leq C_{XQ}(Y') = (C_{XQ}(X'))^h = (Q_0 T)^h$, where $Q_0 = C_Q(X')$. The tori T and $S^{h^{-1}}$ of $Q_0 T$ are conjugate by an element of Q_0, say $S = T^{h_0 h}$ ($h_0 \in Q_0$). Then $X^{h_0 h} = (X'T)^{h_0 h} = Y'S = Y$. Thus X and Y are Q-conjugate, and this completes the proof of Theorem 1.

To conclude this section, we establish an analogue of Theorem 1 for those classical groups which are contained in exceptional groups.

Theorem 3.8 *Let G be a simple exceptional algebraic group over an algebraically closed field of characteristic p, and let C be a simple closed connected subgroup of G of classical type. Suppose that X is a closed connected reductive subgroup of C, and that $p = 0$ or $p > N(X, G)$. When $(X, p) = (B_l, 2)$ or $(C_l, 2)$, assume that $C \neq B_r$ or C_r.*

If X lies in a parabolic subgroup QL of C (with unipotent radical Q and Levi subgroup L), then all closed complements to Q in XQ are Q-conjugate.

Proof. First assume that X is simple, and exclude for the moment the case where $(X, p) = (C_l, 2)$ or $(B_l, 2)$. By [BT], we can embed QL in a parabolic subgroup $Q_1 L_1$ of G such that $Q \leq Q_1$. Suppose that the conclusion of the theorem is false, so that there is a composition factor V of $Q \downarrow X$ such that the semidirect product XV has more than one class of closed complements to V. Since $p = 0$ or $p > N(X, G)$, and V is a composition factor in $Q_1 \downarrow X$, we can apply 3.4 and 3.5 to see that either $(X, p) = (C_3, 3)$ with $V = V_X(\lambda_2)$, or $(X, p) = (A_1, 11 \text{ or } 17)$ with $V = (p-2) \otimes 1^{(p)}$ (note that $(X, p) = (B_r, 2)$ is excluded by hypothesis, and $X = F_4$ does not lie in a classical subgroup of G since any nontrivial representation of F_4 has dimension at least 25). Since Q is the unipotent radical of a parabolic subgroup of the classical group C, it follows that $X = C_3$, $G = E_7$, and QL is an A_5-parabolic in $C = D_6$ (note that $G \neq E_8$ since $N(C_3, E_8) = 3$). Now the conclusion follows using 1.16.

When $p = 2$ and $X = B_l$ or C_l, we have $l \geq 5$ by our hypothesis on p. Also by hypothesis, C is not B_r or C_r, so $C = D_r$ and X lies in a D_{l+1}-parabolic subgroup of C. The conclusion again follows from 1.16.

The result is now proved when X is simple. We extend the argument as before to cover the general case where X is reductive. $\square$

4 Centralizers of reductive subgroups: deduction of Theorem 2

In this section we use Theorem 1 to prove Theorem 2, which states that if X is a closed connected reductive subgroup of the exceptional group G, and $p = 0$ or $p > N(X, G)$, then $C_G(X)^0$ is reductive. The case where X is simple and of rank at least 2 can be read off using the tables which are part of Theorem 5 (to be found in Section 8), but we give a uniform proof for all X here.

At the end of the section we deduce a result on centralizers of non-connected reductive subgroups (see Corollary 4.5).

For inductive purposes in the proof, it is convenient first to prove the following variation of Theorem 2.

Theorem 4.1 *Let G be a simple exceptional algebraic group over an algebraically closed field of characteristic p, and let D be a simple closed connected subgroup of G. Suppose that X is a simple closed connected subgroup of D, and that $p = 0$ or $p > N(X, G)$. Suppose also that $D = G$ if (X, p) is $(B_l, 2)$ or $(C_l, 2)$. Then $C_D(X)^0$ is reductive, and if $p > 0$ then $O_p(C_D(X)) = 1$ (where $O_p(C_D(X))$ denotes the largest normal p-subgroup of $C_D(X)$).*

Notice that this result includes parts (i) and (ii) of Theorem 2 in the case where X is simple. We deduce the general case of Theorem 2 from Theorem 4.1 at the end of the section.

Proof of Theorem 4.1

Let G, D, X be as in 4.1, and let $C = C_D(X)$. For the purposes of proving 4.1, we may take D to be simply connected.

Suppose that 4.1 is false, so that $R_u(C) \neq 1$ or $O_p(C) \neq 1$. Write $U = R_u(C)$ or $O_p(C)$ accordingly, so that $U \neq 1$. By [BT], there is a parabolic subgroup $P_0 = Q_0 L_0$ of D, with unipotent radical Q_0 and Levi subgroup L_0, such that

$$CX \leq P_0, \quad U \leq Q_0.$$

Pick a parabolic subgroup P of P_0, minimal subject to containing X, and write $P = QL$ in the usual way. Then we may assume

$$X \leq P = QL, \quad U \leq Q_0 \leq Q \text{ and } L \leq L_0.$$

Now take the root system $\Sigma(D)$ of D to be based on a maximal torus T of L, and such that P, P_0 are standard parabolics in D. Define $\bar{w}_0$ to be an automorphism of D which normalizes T and induces -1 on $\Sigma(D)$, so that $\bar{w}_0 = w_0$ if $D \neq A_l, D_{2l+1}$ or E_6, and $\bar{w}_0 = w_0 \tau$ if $D = A_l, D_{2l+1}$ or E_6, (where w_0 is the longest element of $W(D)$ and τ is a suitable graph automorphism).

46

By Theorem 1 for D exceptional, or by Theorem 3.8 for D classical, X is Q-conjugate to a subgroup of L'. Replacing X by this conjugate, we have

$$X \leq L', \ CX \leq P_0 \ \text{and} \ U \leq Q_0 \leq Q.$$

Consider now the condition

$(\dagger)$ there exists $l \in L$ such that $\bar{w}_0 l$ normalizes X.

Given $(\dagger)$, we can complete the proof of 4.1 as follows. We have

$$1 \neq |\, C_{Q_0}(X)\,| = |\, C_{Q_0^{\bar{w}_0 l}}(X^{\bar{w}_0 l})\,| = |\, C_{Q_0^-}(X)\,|$$

where $Q_0^- = Q_0^{\bar{w}_0 l}$ is the unipotent radical of the parabolic of D opposite to P_0. But this means that

$$1 \neq C_{Q_0^-}(X) \leq Q_0^- \cap P_0 = 1,$$

which is a contradiction, completing the proof of Theorem 4.1.

Thus we aim to show that $(\dagger)$ holds (under the assumption of the existence of the above nontrivial subgroup U), and prove this in the next three lemmas.

Lemma 4.2 *If $(\dagger)$ fails to hold, then $L' = D_5, A_1 D_5, A_2 D_4, E_6, A_1 E_6$ or D_7.*

Proof. Write $L' = L_1 \ldots L_t$, a direct product of simple groups L_i (it is a direct product since we are taking D to be simply connected). Since $\bar{w}_0$ induces -1 on $\Sigma(D)$, it normalizes each L_i. Let $\pi_i : X \to L_i$ be the projection from X to L_i. We claim that there exists j such that for all $l_j \in L_j$, $\bar{w}_0 l_j$ does not centralize $X \pi_j$. For otherwise, for $1 \leq i \leq t$ there exists $l_i \in L_i$ such that $\bar{w}_0 l_i$ centralizes $X \pi_i$, and then $\bar{w}_0 l_1 \ldots l_t$ centralizes X, so $(\dagger)$ holds, a contradiction.

By the claim just established, the automorphism induced by $\bar{w}_0$ on L_j is not inner, and hence L_j is A_r $(r \geq 2)$, D_{2r+1} $(r \geq 2)$ or E_6. If $L_j = D_{2r+1}$ or E_6 then the conclusion of the lemma holds. So suppose now that $L_j = A_r$ with $r \geq 2$. By the claim in the previous paragraph, $X \pi_j$ is not contained in a symplectic or orthogonal subgroup of A_r (since such subgroups are centralized by suitable elements $\bar{w}_0 l_j$). Hence from Lemma 3.3, we see that either $X \pi_j = L_j$, or $X \pi_j = A_2 < A_r = A_5$. In the latter case, $L' = L_j = A_5$, $X = A_2$ (with $V_{L'}(\lambda_1) \downarrow X = 20$); there is a unique class of irreducible subgroups A_2 in A_5, so $X^{\bar{w}_0} = X^l$ for some $l \in L$ - that is, $(\dagger)$ holds. Therefore $X \pi_j = L_j = A_r$. If $L' = L_j$ then $(\dagger)$ clearly holds. It follows that $L' = A_2 A_2, A_3 A_3$ or $A_2 D_4$. The last case is in the conclusion, so we exclude it from consideration.

Thus $L' = A_r A_r$ $(r = 2$ or $3)$; write $L' = A \times A$ with $A \cong A_r$. Replacing $\bar{w}_0$ by $\bar{w}_0 l$ for suitable $l \in L'$, we may take it that $\bar{w}_0$ induces the automorphism (σ, σ) of L', where σ is a graph automorphism of A of order 2. And replacing X by a suitable L'-conjugate, we may take

$$X = \{(a^{\delta \sigma^i}, a^{\tau \sigma^k}) : a \in A\},$$

for some i, k, where δ, τ are Frobenius morphisms of A which commute with σ. But then X is normalized by $\bar{w}_0$, so (†) holds. $\square$

Lemma 4.3 *Suppose* (†) *fails. Then one of the following holds:*

(i) $L' = D_5$, $X = B_2$ *and* $V_{L'}(\lambda_1) \downarrow X = 02$;

(ii) $L' = D_7$, $X = B_2$, G_2 *or* C_3 *and* $V_{L'}(\lambda_1) \downarrow X = 20, 01$ *or* 010, *respectively;*

(iii) $L' = E_6$ *and* X *is a maximal* A_2 *or* G_2 *in* L';

(iv) $L' = A_2 D_4$ *and* $X = A_2$, *essential in* L'.

Proof. The possibilities for L' are given by 4.2. If $L' = A_2 D_4$ then (iv) holds by the proof of 4.2. Suppose now that L' is D_5, $A_1 D_5$ or D_7. The projection π of X to the factor D_r ($r = 5$ or 7) is a non-parabolic embedding. If $X = A_1$ then the fact that $r = 5$ or 7 implies that $X\pi$ lies in a subgroup $B_s B_{r-s-1}$ of D_r. But this means that $X\pi$ is centralized by $\bar{w}_0 l$ for some $l \in D_r$, and hence (†) holds, a contradiction. Thus X has rank at least 2 (and also $X\pi \not\leq B_s B_{r-s-1}$). From Lemma 3.3 it follows that (i) or (ii) must hold.

Now suppose $L' = E_6$ or $A_1 E_6$, and let π be the projection of X to the factor E_6. By [Se2, Theorem 1], either $X\pi$ lies in a maximal rank subgroup $A_1 A_5$ or A_2^3 of E_6, or $X\pi$ lies in a maximal connected subgroup C_4, F_4, A_2, G_2 or $A_2 G_2$. If $X\pi$ lies in C_4 or F_4, then it is centralized by $\bar{w}_0 l$ for some $l \in L'$, and (†) holds.

Next, assume that $X\pi$ lies in $A_1 A_5$ or A_2^3. Now $\bar{w}_0$ induces graph automorphisms on each factor A_2 or A_5, so if $X = A_1$ then $X\pi$ is centralized by $\bar{w}_0 l$ for some l in $A_1 A_5$ or A_2^3, and hence (†) holds. Therefore $L' = E_6$ and $X = A_2 < A_2^3 < L'$. But now we see as in the last two paragraphs of the proof of 4.2 that (†) again holds.

Now suppose that $X\pi \leq A_2 G_2 < E_6$. The factor G_2 lies in a subgroup F_4 of L', so is centralized by $\bar{w}_0 l$ for some $l \in L'$, and this element $\bar{w}_0 l$ normalizes the factor $A_2 = C_{E_6}(G_2)$. Thus if $X = A_1$, then there exists $m \in A_2$ such that $\bar{w}_0 lm$ centralizes X, giving (†). Consequently $X = A_2$; but then as X embeds in G_2, $X \cong SL_3$ and so X lies in the centralizer in L' of an element of order 3 in G_2, which gives $X < A_2^3$, a case considered before.

We have now established that $X\pi$ lies in A_2 or G_2 (maximal connected subgroups of E_6). Suppose $X = A_1$. By [Se2, p.193 and 4.6], the labelled diagram of $X\pi$ is fixed by a graph automorphism of E_6, and so $X\pi$ is centralized by $\bar{w}_0 l$ for some $l \in L'$, giving (†). If $X\pi = A_2 < G_2$ then as before, X centralizes a 3-element of L', hence lies in A_2^3, a case previously considered. Therefore $X = A_2$ or G_2, and conclusion (iii) of the lemma holds. $\square$

Lemma 4.4 *Cases* (i)-(iv) *of Lemma 4.3 do not occur.*

Proof. Assume that (i),(ii),(iii) or (iv) of 4.3 holds. Observe that $p \neq 2, 3$ by our hypotheses on p. If D is classical, then $L' = A_2 D_4, D_5$ or D_7. In the last two cases, X is irreducible on $V_{L'}(\lambda_1)$, so we have $C_D(X) = C_D(L')$; but $C_D(L')$ is reductive and $O_p(C_D(L')) = 1$, so this contradicts our assumption. And if $L' = A_2 D_4$ then $D = D_7$ or D_8 and $V_D(\lambda_1) \downarrow X = 10/01/11/00^a$ ($a = 0$ or 2); then again $O_p(C_D(X)) = 1$, a contradiction. Therefore D is exceptional.

If $L' = A_2 D_4$ then $D = E_8$ and $Z(X) = \langle t \rangle \cong Z_3$, so $C_D(X) \leq C_D(t) = A_2 E_6$. Let π be the projection of X to the D_4 factor of L'. Then $C_D(X) = C_{E_6}(X\pi)$. Hence we may replace D by E_6, L' by D_4 and X by $X\pi$ in this case.

Now D is exceptional, of type E_6, E_7 or E_8. Using 2.1, we work out the composition factors of $L(D) \downarrow L'$:

$$
\begin{aligned}
L(E_8) \downarrow D_5 &= L(D_5)/\lambda_1^6/\lambda_4^4/\lambda_5^4/0^{15}, \\
L(E_7) \downarrow D_5 &= L(D_5)/\lambda_1^2/\lambda_4^2/\lambda_5^2/0^4, \\
L(E_6) \downarrow D_5 &= L(D_5)/\lambda_4/\lambda_5/0, \\
L(E_8) \downarrow D_7 &= L(D_7)/\lambda_1^2/\lambda_6/\lambda_7/0, \\
L(E_8) \downarrow E_6 &= L(E_6)/\lambda_1^3/\lambda_6^3/0^8, \\
L(E_7) \downarrow E_6 &= L(E_6)/\lambda_1/\lambda_6/0, \\
L(E_6) \downarrow D_4 &= L(D_4)/\lambda_1^2/\lambda_3^2/\lambda_4^2/0^2.
\end{aligned}
$$

The group $C_D(L')$ is a reductive subgroup, with semisimple part corresponding to the subsystem orthogonal to the root system of L'. Working out these subgroups, we find that

$$\dim C_D(L') = \dim C_{L(D)}(L') = \text{no. of trivial comp. factors in } L(D) \downarrow L'.$$

For any nontrivial composition factor $V_{L'}(\lambda)$ of $L(D) \downarrow L'$, the restriction $V_{L'}(\lambda) \downarrow X$ is given by 2.4 and 2.5 when $L' = E_6$, and by 2.10 and 2.12 when $L' = D_5$ or D_7. These results imply that X has no trivial composition factors on such modules $V_{L'}(\lambda)$, and so we deduce that $L(D) \downarrow X$ has the same number of trivial composition factors as $L(D) \downarrow L'$. Hence $C^0 = C_D(X)^0 = C_D(L')^0$, which is reductive. Thus $R_u(C) = 1$, which means that $U = O_p(C)$ is a finite p-group. Then $[U, C^0] = 1$, so $X \times U$ lies in $C_D(C^0)$. But $C_D(C^0) = L'$, so this contradicts the fact that X is non-parabolically embedded in L'. This final contradiction proves the lemma. $\square$

Lemmas 4.2 - 4.4 establish the condition (†), from which we have seen that Theorem 4.1 follows. Thus the proof of 4.1 is now complete.

We now deduce parts (i) and (ii) of Theorem 2 from 4.1. Let G be a simple exceptional algebraic group in characteristic p, and let D be a simple closed connected subgroup of G. Suppose X is a closed connected reductive subgroup of D, and $p = 0$ or $p > N(X, G)$. Then $X = X'Z$, where $Z = Z(X)^0$ and $X' = X_1 \ldots X_t$, a commuting product of simple groups X_i.

If $p = 2$ and some factor X_i is B_l or C_l, then by our assumption on p we must have $l \geq 4$ (and $l \geq 5$ if $G \neq F_4$). This forces $t = 1$, since any further factor would have rank at most 3, and hence be excluded by our assumptions on p. Now the conclusions of Theorem 2(i,ii) follow from 4.1. Thus we may suppose that if $p = 2$, then no factor X_i is B_l or C_l.

We now argue by induction on t that $C_D(X)$ is reductive and $O_p(C_D(X)) = 1$. This is true for $t = 1$ by 4.1, so assume $t \geq 2$. Write $E = C_D(X_1)^0$. By 4.1, E is reductive, so $E = E'Z_1$ where $Z_1 = Z(E)^0$ and $E' = E_1 \ldots E_r$, a commuting product of simple groups E_i. By induction, for each i the group $F_i = C_{E_i}(X_2 \ldots X_t Z)$ is reductive. Then $C_D(X)^0 = \prod F_i^0 Z_1$ is reductive. Furthermore, our assumptions on p force the group $U = O_p(C_D(X))$ to induce inner automorphisms on E'. By induction, U centralizes E'. Now $C_D(X_1)$ normalizes $C_D(X_1 E')$, which contains U and lies in $C_D(X)$. Hence $U \leq O_p(C_D(X_1))$. But $O_p(C_D(X_1)) = 1$ by 4.1, so $U = O_p(C_D(X)) = 1$.

In particular, taking $D = G$ we conclude that $C_G(X)$ is reductive, and $O_p(C_G(X)) = 1$, proving parts (i) and (ii) of Theorem 2.

To complete the proof of Theorem 2, it remains to establish part (iii) - that when X is semisimple the rank of the reductive group $C_G(X)$ is equal to the maximal co-rank among subsystem subgroups of G containing X. Let Y be a subsystem subgroup containing X of minimal rank, and let T be a maximal torus of $C_G(Y)$. It will suffice to show that T is also a maximal torus of $C_G(X)$.

Let $L = C_G(T)$, a Levi subgroup of G. Then $X \leq Y \leq L'$, so Y and L' have the same rank, by choice of Y. If T_0 is a torus of $C_G(X)$ which properly contains T, then $X \leq C_{L'}(T_0)$, which is a proper Levi subgroup of L'; hence X lies in a subsystem subgroup of rank smaller than that of L', which is a contradiction. It follows that T is a maximal torus of $C_G(X)$.

This completes the proof of Theorem 2.

We conclude this section with the following result on centralizers of non-connected subgroups of G.

Corollary 4.5 *Let G be an exceptional simple algebraic group in characteristic p, and suppose that R is a subgroup of G such that R^0 is reductive and R/R^0 is a p'-group. Assume that either $p = 0$ or $p > N(R^0, G)$, and also that p is a good prime for G.*

Then $C_G(R)$ is reductive, and $O_p(C_G(R)) = 1$.

For the proof of this result we require the following fact concerning p'-automorphisɪ of semisimple algebraic groups.

Lemma 4.6 *Let A be a semisimple algebraic group in characteristic $p > 0$.*

(i) If x is an automorphism of A of finite p'-order, then $C_A(x)$ is reductive, and $C_A(x)/C_A(x)^0$ is a p'-group.

(ii) Let R be a finite p'-subgroup of A. Then $C_A(R)$ is reductive and $O_p(C_A(R)) = 1$.

Proof. (i) Using induction on the order of x, it suffices to establish the result when x has prime order, r say. If x induces a nontrivial permutation on a set $N_1, \ldots, N_r$ of simple components N_i of A, then $C_{N_1 \ldots N_r}(x) \cong N_1$. Hence we may assume that x fixes every simple component of A.

Thus it suffices to establish the result for the case where A is simple and x is of prime order r. When x is an inner automorphism, the result follows from [SS, II, 4.4]. Otherwise, A is of type A_l, D_l, D_4 or E_6, and x is a graph automorphism of order $r = 2, 2, 3$ or 2, respectively. The possibilities for $C_A(x)$ are well known for these cases (see [CLSS, 2.7] for example): for $A = A_l$ they are $C_{(l+1)/2}, D_{(l+1)/2}$ or $B_{l/2}$; for $A = D_l, r = 2$ they are $B_s B_{l-s-1}$ ($s < l$); for $A = D_4, r = 3$ they are G_2, A_2; and for $A = E_6$ they are F_4, C_4. Part (i) follows.

(ii) Suppose (ii) is false, and let $U = R_u(C_A(R))$ or $O_p(C_A(R))$ with $U \neq 1$. By [BT], there is a parabolic subgroup $P = QL$ such that $RC_A(R) \leq P$ and $U \leq Q$. As R is a p'-group, R lies in a Q-conjugate of L, so may may assume that $R \leq L$. Let Q^- be the unipotent radical of the parabolic opposite to P. We claim that $C_{Q^-}(R) \neq 1$, which is a contradiction; this will establish (ii). To see this, observe as usual that Q, Q^- have filtrations by high weight modules for L; the modules appearing in Q^- are the duals of those in Q. Hence R has a fixed point in some L-invariant section V/W of Q^-; say R fixes Wv (where $v \in V - W$). Then $(WR)^v = WR$. As WR has only one class of complements to W, it follows that $R^v = R^w$ for some $w \in W$. Then vw^{-1} normalizes R, hence centralizes R, so $vw^{-1} \in C_{Q^-}(R)$ and $C_{Q^-}(R) \neq 1$, as claimed. $\square$

We now embark upon the proof of Corollary 4.5. Write $X = R^0$, so that X is connected and reductive. By 4.6(b), we may assume that $X \neq 1$. By Theorem 2, $C_G(X)$ is reductive and $O_p(C_G(X)) = 1$. Thus

$$C_G(X)^0 = AT,$$

where A is the semisimple group $(C_G(X)^0)'$ and $T = Z(C_G(X)^0)^0$. Then $C_G(R)^0 = C_A(R)^0 C_T(R)^0$. If the group of outer automorphisms of A induced by R is solvable, then $C_G(R)^0$ is reductive by 4.6. Assume now that this group of outer automorphisms is non-solvable. Then $A = A_1^k$ for some $k \geq 5$. Hence A contains a torus S of rank at least 5, and $C_G(S)$ is a Levi subgroup L of semisimple rank at most $\text{rank}(G) - k \leq 3$. Then $X \leq L$, so $C_G(L)^0 \leq A$. But it is easy to check that for such a Levi subgroup L, $C_G(L)^0$ must have a simple factor of rank bigger than 1, which is a contradiction as $A = A_1^k$. This contradiction establishes that $C_G(R)^0$ is reductive.

It remains to show that $O_p(C_G(R)) = 1$ (when $p > 0$). Let $U = O_p(C_G(R))$, and suppose that $U \neq 1$. Write $C = C_{C_G(X)}(A)$. Then $T \leq C$ and C/T is finite.

We claim that $C_G(X)/AC$ is a finite p'-group. To see this, let $A = N_1 \ldots N_k$, a commuting product of simple factors N_i. Then $S = C_G(X)/AC$ acts on the set of factors $N_1, \ldots, N_k$, and the kernel P of this action induces trivial or graph automorphisms on each factor, hence is a $\{2,3\}$-group. Suppose that p divides $|S|$. Since p is a good prime, $p > 3$, and so p divides $|S/P|$. Hence $k \geq p$. Certainly k is smaller than the rank of G, so we conclude that $p = 5$ if $G \neq E_8$, and $p = 7$ if $G = E_8$. It follows that A contains a normal subgroup N which is the product of p factors A_1. Since $p > N(X, G)$, it is not possible that $X = A_1$, and hence as $X \leq C_G(N)$, it must be the case that $p = 5$, $G = E_7$ and X has rank 2. As $p > N(X, G)$, we have $X = B_2$. But this means that $A_1^5 B_2 < E_7$, whereas E_7 has no maximal rank subgroup of this form. This contradiction establishes the claim that $S = C_G(X)/AC$ is a p'-group.

It follows that $U \leq AC$. Also $C_A(R)$ normalizes the projection of U in A, so since $O_p(C_A(R)) = 1$ by 4.6, we have $U \leq C$ in fact.

We claim next that U centralizes T. Recall that $T \leq C$ and C/T is finite. Suppose that U does not centralize T. Then p divides $|W(G)|$. As p is a good prime, this forces $p = 7$ if $G = E_8$, and $p = 5$ otherwise. If $p = 7$ then T has rank at least 6: for any element of order 7 in $W(G)$ lies in $W(A_6)$ for some subsystem A_6, and normalizes (without centralizing) only tori of rank 6 or more. Since $X \leq C_G(T)$ this means that X lies in a Levi subgroup of semisimple rank at most 2. Because of our assumption that $p > N(X, G)$, we have $X \neq A_1$, and hence X is a Levi subgroup A_2 of $G = E_8$. But then $C_G(X) = E_6$ which does not contain T as a normal subgroup. This rules out the case where $p = 7$. Similar considerations deal with the case where $p = 5$. This proves our claim that U centralizes T.

We have now established that U centralizes $AT = C_G(X)^0$. Let $D = C_G(T)'$, a connected semisimple group containing XAU. Then the double centralizer $C_D(C_D(U))$ lies in $C_D(XA)$, which is finite. However, p is a good prime, so for $u \in U$ of order p, $Z(C_D(u))$ is infinite (see the argument of [LS1, p.358]). But $Z(C_D(u)) \leq C_D(C_D(u)) \leq C_D(C_D(U))$, so $C_D(C_D(U))$ is infinite, which is a contradiction. This completes the proof of the corollary. $\square$

5 Embedding of reductive subgroups: proofs of Theorems 5 and 7

In this section we prove Theorems 5 and 7. These are both short deductions from Theorem 1, apart from the calculations which justify the information given in Tables 8.1-8.7; these calculations form the bulk of this section.

Let G be an exceptional simple algebraic group over an algebraically closed field of characteristic p, and let X be a semisimple closed connected proper subgroup of G. Assume that $p = 0$ or $p > N(X, G)$.

Choose a subsystem subgroup Y of G such that $X \leq Y$ and Y is minimal with this property. Then X lies in no proper subsystem subgroup of Y. By Theorem 1 and Theorem 3.8, X also lies in no proper parabolic subgroup of Y.

Lemma 5.1 *We have $C_Y(X) = Z(Y)$.*

Proof. Suppose false. We may choose a non-identity element $u \in C_Y(X) - Z(Y)$ such that u is unipotent or semisimple. Then $C_Y(u)$ lies in a proper parabolic or subsystem subgroup of Y. Since $X \leq C_Y(u)$, this is a contradiction. $\square$

Lemma 5.2 *Suppose that $X \leq H \leq Y$, where H is a semisimple subgroup of Y. Let D be a simple factor of H, and let X_0 be the projection of X in D (as defined in the Introduction).*

(i) If D is of classical type or of type G_2, then X_0 is essentially embedded in D.

(ii) If D is of type F_4, then either X_0 is essentially embedded in D, or there is a maximal connected subgroup $Z = A_1 G_2 \, (p \neq 2)$ or $G_2 \, (p = 7)$ of D such that X_0 is essentially embedded in Z.

Proof. Assume first that D is classical of type A_n, B_n or C_n, and let V be the usual module for the simply connected cover of D. Observe that X_0 lies in no proper parabolic subgroup of D (as otherwise X would lie in a proper parabolic subgroup of Y). Suppose X_0 lifts to a reducible subgroup on V; then $D = B_n$ or C_n and X_0 lies in a reducible maximal rank subgroup M of D. By 5.1, X_0 centralizes no simple factor of M, so X_0 projects nontrivially to each factor. Consequently X_0 has a simple factor of rank at most $\frac{n}{2}$, and so $p \neq 2$ by our assumptions on p. But this means that $Z(M) \neq 1$; since $Z(M) \leq C_Y(X)$, this contradicts 5.1. Thus X_0 lifts to an irreducible subgroup on V. Similarly, when $D = D_n$, (the lift of) X_0 is either irreducible, or contained in a subgroup $B_r B_{n-r-1}$ of D, irreducible in each factor with inequivalent representations, or contained in a subgroup D_{n-1}. In the last case, as before when $p \neq 2$ we have a contradiction by 5.1. Therefore $p = 2$, whence $n = 4$ and $G = F_4$ by our assumptions on p. But then $X = A_3$ is a subsystem subgroup of G, contrary to our choice of Y. Thus this case cannot occur. We have now established that when D is classical, X_0 is essentially embedded in D.

Now suppose that $D = G_2$ or F_4, and assume that $X_0 \neq D$. Choose a maximal connected subgroup M of D containing X_0. Then M is not parabolic. And if M is of maximal rank, then by our assumptions on p we have $Z(M) \neq 1$, again contrary to 5.1. Therefore M is a maximal connected subgroup of D not containing a maximal torus. By [Se2, Theorem 1], the possibilities for (M, D) are as follows:

$$(A_1, G_2)\,(p = 0 \text{ or } p \geq 7), \quad (A_1, F_4)\,(p = 0 \text{ or } p \geq 13),$$

$$(G_2, F_4)\,(p = 7), \quad (A_1 G_2, F_4)\,(p \neq 2).$$

Therefore if $D = G_2$ then $X_0 = M$ and (i) holds. Now consider $D = F_4$. If $M = A_1$ then $X_0 = M$; and in the other two cases, the preceding argument shows that X_0 is essential in M. This proves (ii). □

If Y is a commuting product of classical simple groups, then by 5.2, conclusion (i) of Theorem 5 or Theorem 7 holds. Thus we suppose now that Y has a factor which is an exceptional group. If this factor is of type G_2 or F_4, then $Y = G = G_2$ or F_4 and 5.2 shows that Theorem 5(i) or Theorem 7 holds. Thus we may assume that Y has a factor E_r with $r = 6, 7$ or 8; hence $Y = E_6, A_1 E_6, A_2 E_6, E_7, A_1 E_7$ or E_8. Since $C_Y(X) = Z(Y)$ by 5.1,, X projects nontrivially to each factor of Y. Also, if $X = Y$ then Theorem 5(i) or Theorem 7(ii) holds, so assume that $X < Y$.

At this point we consider separately the cases where X has no factor A_1 (Theorem 5), and where X has a factor A_1 (Theorem 7).

Proof of Theorem 5

Suppose that all the simple factors of X have rank at least 2. Then either Y is simple or $Y = A_2 E_6$. For the moment, exclude the case where $Y = A_2 E_6$. Pick a maximal connected subgroup M of Y containing X. Then M is not a parabolic or subsystem subgroup of Y, and also M has no factor A_1 (for otherwise $C_Y(X) \geq A_1$, since X has no factor A_1). Hence by [Se2, Theorem 1], one of the following holds (note that the prime restrictions of [Se2] hold, by our assumptions on p):

$$Y = E_6, \quad M = A_2, G_2, F_4, C_4 \text{ or } A_2 G_2;$$
$$Y = E_7, \quad M = A_2 \text{ or } G_2 C_3;$$
$$Y = E_8, \quad M = B_2 \text{ or } G_2 F_4.$$

Suppose that M has a factor G_2. Then by 5.2, the projection of X in this factor is equal to G_2. Therefore, if $Y = E_6$ then $X = M$ and Theorem 5(i) holds. If $Y = E_7$ or E_8, then $M = G_2 C_3$ or $G_2 F_4$, and $p = 0$ or $p > 7$ (since $N(G_2, E_7) = N(G_2, E_8) = 7$, as defined in the Introduction). Now application of 5.2 forces $X = M$, as in Theorem 5(i).

Thus we may assume that M is simple and $M \neq G_2$. If M is A_2 or B_2, then $X = M$ and (i) of Theorem 5 holds. The only remaining cases are $Y = E_6$, $M = F_4$ or C_4. If $M = C_4$ then $p \neq 2$ (see [Se2, Theorem 1]), and 5.2 implies that X is an irreducible subgroup of M; and now 2.8 forces $X = M$, as in Theorem 5(i). If $M = F_4$ then by 5.2, either $X = M$ or $p = 7$ and X is a maximal subgroup G_2 of M. Hence (i) or (ii) of Theorem 5 holds.

Finally, suppose that $Y = A_2 E_6$ (excluded from consideration earlier). The projection of X in the factor E_6 of Y lies in no parabolic or subsystem subgroup, so we see as above that this projection is either E_6 or a maximal connected subgroup of E_6 (note that it is not a maximal G_2 in a subgroup F_4 with $p = 7$, as $N(G_2, E_8) = 7$); hence conclusion (i) of Theorem 5 holds again.

We have now established that when all factors of X have rank at least 2, conclusion (i) or (ii) of Theorem 5 holds. To complete the proof of Theorem 5, it remains to justify the information given in Tables 8.1-8.5 for simple subgroups X of rank at least 2. We postpone this until after the proof of Theorem 7.

Proof of Theorem 7

Suppose X has a factor A_1. If $G = G_2$ then $Y = G$ and X is a maximal connected subgroup A_1 of G, as in (iii) of Theorem 7. Thus we suppose that $G \neq G_2$. Note that our assumptions on p include all the prime restrictions in [Se2, Theorem 1].

Assume first that X does not lie in $N_G(F_4)$ for any subgroup F_4 of G. Let Y_0 be the factor E_6, E_7 or E_8 of Y. If $Y_0 \leq X$ then $X = Y_1 Y_0$ as in conclusion (ii)(a) of Theorem 7, so assume $Y_0 \not\leq X$. Then the projection X_0 of X in Y_0 lies in a maximal connected subgroup Z of Y_0, where Z is not a parabolic or subsystem subgroup of Y_0. From [Se2], we deduce that one of the following holds:

$$Y_0 = E_6, \quad Z = A_2, G_2, C_4, \text{ or } A_2 G_2;$$
$$Y_0 = E_7, \quad Z = A_1, A_2, A_1 A_1, A_1 G_2 \text{ or } G_2 C_3;$$
$$Y_0 = E_8, \quad Z = A_1, B_2 \text{ or } A_1 A_2.$$

By 5.2, X_0 is essentially embedded in Z. The projection of X in $C_Y(Y_0)$ (which is 1, T_1, A_1 or A_2) must also be essential in $C_Y(Y_0)$. Hence the conclusion (ii)(b) of Theorem 7 holds.

Finally, suppose that X lies in $N_G(F_4)$. We have proved above that any subgroup F_4 of G satisfies conclusion (i) or (ii) of Theorem 5. Consequently, if $G = E_6, E_7$ or E_8, then every subgroup F_4 of G lies in a subsystem subgroup E_6 of G. It follows that there is just one conjugacy class of subgroups F_4 in G. From [LS2, Section 4], we see that $N_G(F_4) = F_4$, F_4, $A_1 F_4$ or $G_2 F_4$, according as $G = F_4$, E_6, E_7 or E_8. Now conclusion (ii) of Theorem 7 follows from 5.2.

This completes the proof of Theorem 7.

Proof of Tables 8.1-8.7

The rest of the section is devoted to proving all the information given in the tables in Section 8. Thus, for each simple exceptional algebraic group G, we calculate all conjugacy classes and centralizers in G of simple subgroups X of rank at least 2, the minimal subsystem subgroups of G containing X, and the composition factors of $L(G) \downarrow X$. (The information in the last two tables, 8.6 and 8.7, giving the restrictions $V_{56} \downarrow X$, $V_{27} \downarrow X$ for $G = E_7$, E_6 respectively, follows easily from 2.3 and 2.5, using the (known) minimal subsystem group containing X.)

We carry out these calculations for $G = E_8$, proving the information in Table 8.1. After this, we show how to deduce from this table the information in all the other tables.

Suppose then that $G = E_8$ and X is a proper closed connected simple subroup of G of rank at least 2. Assume that $p = 0$ or $p > N(X, G)$. Let Y be a subsystem subgroup of G, minimal subject to containing X. By Theorem 5, with the one exception described in Theorem 5(ii), X is essentially embedded in Y. Write $\Sigma(G)$ for the root system of G. For convenience in determining the possibilities for Y, we state a lemma which is immediate from the definition of essential embedding.

Lemma 5.3 *Suppose that X is essentially embedded in Y, let Y_0 be a simple factor of Y and let X_0 be the projection of X in Y_0. Then one of the following holds:*

(i) Y_0 is classical, with usual module V, and X_0 lifts to an irreducible subgroup on V; moreover, the high weight of $V \downarrow X_0$ is given by 2.8;

(ii) $Y_0 = D_n$ and $X_0 < B_r B_{n-r-1}$ for some r, irreducible in each factor with high weights given by 2.8;

(iii) Y_0 is exceptional and either $X_0 = Y_0$ or X_0 is a maximal connected subgroup of Y_0 not containing a maximal torus, as determined by [Se2, Theorem 1] (and listed in the table in the Remark after Theorem 7).

Consider first the case where $\operatorname{rank}(X) \geq 5$. Observe that X is not C_l or B_8, as C_5, B_8 do not lie in any subsystem group Y as in (i) or (ii) of Theorem 5. Hence X is A_l, D_l ($5 \leq l \leq 8$), B_l ($5 \leq l \leq 7$), E_6 or E_7, as in Table 8.1. When $X \neq B_l$, it is clear that $Y = X$, since there are no other possible subsystem groups which contain X minimally. Thus X is a subsystem group corresponding to a closed subsystem Δ of $\Sigma(G)$. Then $C_G(X)^0$ is the subsystem group corresponding to the orthogonal subsystem $\Delta^{\perp}$, and so we see that $C_G(X)^0$ is as in Table 8.1 when $X \neq B_l$. Note that E_8 contains two non-conjugate subsystems of type A_7, one a Levi subsystem and the other contained in a subsystem E_7; we denote these by A_7 and A_7', respectively.

Now suppose $X = B_l$ ($5 \leq l \leq 7$). Then $Y = D_{l+1}$ and Y lies in a subsystem group D_8. We have $C_{D_8}(X)^0 = B_2$, A_1 or 1, according as $l = 5, 6$ or 7, respectively. By Theorem 2(iii), $C_G(X)$ has rank $8 - l - 1$. Hence $C_G(X)^0 = C_{D_8}(X)^0$, as in Table 8.1.

We now calculate $L(G) \downarrow X$ for rank $(X) \geq 5$. If $X = E_6$ or E_7, this follows from 2.1. Otherwise, X lies in D_8 or A_8, and $L(G) \downarrow X$ can be found by restricting from $L(G) \downarrow D_8$ or $L(G) \downarrow A_8$ (given in 2.1), also using 2.6 to restrict from $L(G) \downarrow D_8$ to $L(G) \downarrow A_7'$ (note that A_7 lies in both A_8 and D_8, while A_7' lies only in D_8). We have now established all the information in Table 8.1 when rank $(X) \geq 5$.

For rank $(X) \leq 4$, we deal separately with each possibility for X. Our method is as follows:

1. list all subsystem groups Y which are minimal subject to containing a subgroup isomorphic to X,

2. calculate $L(G) \downarrow Y$, hence $L(G) \downarrow X$,

3. determine the G-conjugacy classes of subgroups X, using (2),

4. find $C_G(X)^0$.

For (1), the possibilities for Y are given by 5.3. (To list the possibilities, it is convenient to use the table of all subsystems of $\Sigma(G)$ given in [Ca, Table 11].) For (2), the restriction $L(G) \downarrow Y$ is usually easy to deduce from 2.1 and 2.2. As for (4), we usually first use (2) to calculate the number t of trivial composition factors in $L(G) \downarrow X$. Then clearly $\dim C_G(X) \leq t$. Often we can produce a connected subgroup C of $C_G(X)$ of dimension t, and hence conclude that $C_G(X)^0 = C$.

Case $X = A_4$

The only possibilities for Y (a subsystem group containing X minimally) are A_4 and A_4A_4. When $Y = A_4$ we have $X = Y$ and $C_G(X)^0 = A_4$; also the restriction $L(G) \downarrow X$ is easily obtained by restricting from $L(G) \downarrow A_4A_4$ (given in 2.1).

Now suppose that $Y = A_4A_4$. Here $N_G(Y)/Y = \langle w \rangle \cong Z_4$ (see [Ca, Table 11]), where w interchanges the factors A_4, and w^2 induces an element in the coset of a graph automorphism on each factor. For some fixed copy A of A_4, write the elements of Y as $a.b\,(a, b \in A)$. Then we may assume that

$$X = \{a.a^\phi : a \in A\}$$

for some morphism ϕ of A. If τ is a graph automorphism of A, and σ_p is a pth power field automorphism, then by [St, Chapter 12] we have $\phi = \alpha\tau^i\sigma_p^j$, where α is an inner automorphism, $i = 0$ or 1, and $j \geq 0$.

Assume first that $j = 0$. Then X is Y-conjugate to X_1 or X_2, where

$$X_1 = \{a.a : a \in A\}, \quad X_2 = \{a.a^\tau : a \in A\}.$$

We claim that w interchanges these two Y-classes. For suppose not, and pick an element u in the Y-coset of w normalizing X_1. Then u^2 induces an inner automorphism on X_1, so $C_Y(t)^0 = X_1$ for some involution t in the coset of w^2. But w^2

induces an element in the coset of a graph automorphism on each factor A_4 of Y, so $C_Y(t) = B_2 B_2$. This contradiction establishes the claim. We conclude that there is just one G-class of subgroups $X = A_4$ in this case ($j = 0$). Restricting from $L(G) \downarrow A_4 A_4$ given in 2.1, we find that

$$L(G) \downarrow X = L(A_4)^2/\lambda_1 \otimes \lambda_2/\lambda_2 \otimes \lambda_4/\lambda_3 \otimes \lambda_1/\lambda_4 \otimes \lambda_3.$$

The composition factors of the tensor products $\lambda_i \otimes \lambda_j$ are given by 2.14, and so we see that $L(G) \downarrow X$ is as in Table 8.1. Finally, since X lies in no Levi subgroup and in no parabolic, $C_G(X)^0 = 1$.

Assume now that $j \geq 1$, so that ϕ involves the Frobenius morphism $\sigma_p^j = \sigma_q$ (where $q = p^j$). Then X is Y-conjugate to $\{a.a^{\sigma_q} : a \in A\}$ or $\{a.a^{\tau\sigma_q} : a \in A\}$. These two subgroups are not G-conjugate, as they have different actions on $L(G)$ (the restrictions of $L(G)$ to these subgroups follow from $L(G) \downarrow A_4 A_4$, given in 2.1). As before, $C_G(X)^0 = 1$.

Case $X = B_4 \, (p \neq 2)$

Note that $p \neq 2$ by hypothesis (since $N(B_4, E_8) = 2$, as defined in the Introduction). The minimal subsystem groups Y containing B_4 are D_5, A_8 and D_8 (see 2.8).

If $Y = D_5$ or A_8, then $L(G) \downarrow Y$ is as in Table 8.1, and we obtain $L(G) \downarrow X$ by restriction from this, noting that for $Y = A_8$, $L(Y) \downarrow B_4 = L(B_4)/2\lambda_1$ and $V_Y(\lambda_3) \downarrow B_4 = V_{B_4}(\lambda_3)$ (see 2.10). When $Y = A_8$, the number t of trivial composition factors in $L(G) \downarrow X$ is 0, so $C_G(X)^0 = 1$; and when $Y = D_5$, $t = 21$, so $C_G(X)^0 = C_{D_8}(X)^0 = B_3$. Observe also that if $Y = A_8$ then X centralizes an involution $t \in N_G(Y) = A_8.2$, so X lies in $C_G(t) = D_8$. This D_8 is a minimal subsystem subgroup containing X, since otherwise $C_{D_8}(X)^0 \neq 1$.

Now consider $Y = D_8$. Here $V_Y(\lambda_1) \downarrow X = V_X(\lambda_4)$ (see 2.8). There are two Y-classes of such subgroups $X = B_4$ (their images in the adjoint group $Y/Z(Y)$ are interchanged by a graph automorphism of Y). Since Y is a half-spin group, one of the classes contains a group $X_1 \cong \mathrm{Spin}_9$, and the other a group $X_2 \cong SO_9$. To calculate $L(G) \downarrow X_i$, note first that by 2.1 and 1.1,

$$L(G) \downarrow D_8 = L(D_8)/\lambda_7 = \lambda_2/\lambda_7.$$

Now $L(D_8) \downarrow X_i = L(B_4)/\lambda_3$ for $i = 1$ or 2, by 2.10; and by 2.12, $V_Y(\lambda_7) \downarrow X_1 = \lambda_1 + \lambda_4$, $V_Y(\lambda_7) \downarrow X_2 = \lambda_3/2\lambda_1$. Hence $L(G) \downarrow X_i$ is as in Table 8.1, and X_2 lies in A_8 (see the last remark in the previous paragraph). Since $t = 0$, $C_G(X_1)^0 = 1$.

Case $X = C_4 \, (p \neq 2)$

The minimal subsystem groups Y containing C_4 are E_6, A_7 and A_7'. Moreover, if $X < E_6$ then also $X < A_7'$ by the proof of [LS2, 4.9]. We obtain $L(G) \downarrow X$ by restricting from $L(G) \downarrow A_7$ and $L(G) \downarrow A_7'$, noting that $V_{A_7}(\lambda_i) \downarrow C_4 \, (1 \leq i \leq 4)$ is given in 2.10 and 2.11. When $Y = A_7$, we have $t = 3$ (where as always, t is the number of trivial composition factors in $L(G) \downarrow X$), so $C_G(X)^0 = C_{D_8}(X)^0 = A_1$ (from $C_1 \otimes C_4 < D_8$). And when $Y = E_6$, $C_G(X)^0 \geq C_G(Y) = A_2$; since XA_2 lies in no parabolic subgroup, $C_G(X)^0 = A_2$.

Case $X = D_4 \, (p \neq 2)$

Here Y is D_4, $D_4 D_4$, A_7 or A_7'. If $Y = A_7$ or A_7' then $Y < D_8$, and $V_{D_8}(\lambda_1) \downarrow Y = V_1 \oplus V_2$ where V_1, V_2 are totally singular subspaces of dimension 8. Moreover, $V_1 \downarrow X \cong V_2 \downarrow X$, and so $X < D_4 D_4 < D_8$. Thus we need consider only $Y = D_4$ or $D_4 D_4$. When $Y = D_4$, $C_G(X)^0 = D_4$ and $L(G) \downarrow X$ is easily obtained from $L(G) \downarrow D_5$.

Now suppose $Y = D_4 D_4$. By [Ca, Table 11], $N_G(Y)/Y \cong S_3 \times Z_2$, where the S_3 induces graph automorphisms on both factors D_4, and the Z_2 interchanges the factors. Moreover, $Z(Y) \cong Z_2^2$ and $Z(Y) = Z(D_4)$ for both factors D_4. For a fixed copy D of the simply connected group D_4, write the elements of Y as $a.b \, (a, b \in D)$. Then we may assume that
$$X = \{a.a^\phi : a \in D\}$$
for some morphism ϕ of D.

First assume that ϕ involves no Frobenius morphism of D. Then X is $N_G(Y)$-conjugate to X_1, X_2 or X_3, where

$$X_1 = \{a.a : a \in D\}, \quad X_2 = \{a.a^\iota : a \in D\}, \quad X_3 = \{a.a^\tau : a \in D\}$$

and ι, τ are graph automorphisms of D of order 2,3 respectively. Moreover,

$$Z(X_1) = 1, \quad Z(X_2) \cong Z_2, \quad Z(X_3) \cong Z_2^2.$$

Since $Z(A_7) \cong Z_8$ and $Z(A_7') \cong Z_4$, we see that A_7, A_7' contain subgroups D_4 with centres of order 2,1 respectively. We deduce that $X_1 < A_7', X_2 < A_7$ while X_3 lies in no subgroup of type A_7. The restriction $L(G) \downarrow D_4 D_4$ is given in 2.1, and we see from this that

$$
\begin{aligned}
L(G) \downarrow X_1 &= L(D_4)^2/\lambda_1 \otimes \lambda_1/\lambda_3 \otimes \lambda_3/\lambda_4 \otimes \lambda_4 \\
L(G) \downarrow X_2 &= L(D_4)^2/\lambda_1 \otimes \lambda_1/\lambda_3 \otimes \lambda_4/\lambda_4 \otimes \lambda_3 \\
L(G) \downarrow X_3 &= L(D_4)^2/\lambda_1 \otimes \lambda_3/\lambda_3 \otimes \lambda_4/\lambda_4 \otimes \lambda_1.
\end{aligned}
$$

The composition factors of $L(G) \downarrow X_i$ can now be read off from 2.14. From these restrictions we see that $C_G(X_3)^0 = 1$ and $C_G(X_2)^0 = T_1$; and from $L(G) \downarrow X_1$, we have $t = 3$, hence $C_G(X_1)^0 = C_G(A_7')^0 = A_1$.

Now suppose that ϕ involves a Frobenius q-power morphism σ_q. As above, there are 3 $N_G(Y)$-classes of subgroups X, with representatives $\{a.a^{\gamma \sigma_q} : a \in D\}$, where $\gamma = 1, \iota$ or τ. The restrictions of $L(G)$ to these subgroups follow from $L(G) \downarrow D_4 D_4$, given in 2.1; all 3 restrictions have different sets of composition factors, so the 3 subgroups are not G-conjugate; and $C_G(X)^0 = 1$ in all cases.

Case $X = F_4$

In this case, $Y = E_6$. We know that $L(G) \downarrow Y = L(E_6)/\lambda_1^3/\lambda_6^3/0^8$. The restrictions $L(E_6) \downarrow F_4$ and $V_{E_6}(\lambda_1) \downarrow F_4$ are given in 2.4 and 2.5. Hence $L(G) \downarrow X$ is as in Table 8.1. Finally, $C_G(X)^0 = G_2$ by [LS2, 4.6].

Case $X = A_3\,(p \neq 2)$

For a subsystem subgroup A_3, we have $C_G(A_3) = D_5$, which contains 2 classes of subgroups of type A_3. Thus there are two G-classes of subsystem subgroups of type A_3^2, which we shall denote by A_3A_3 and A_3A_3': we take A_3A_3 to lie in the Levi subsystem group A_7, and A_3A_3' to lie in A_7'. Thus $|\,Z(A_3A_3)\,| = 16$ and $|\,Z(A_3A_3')\,| = 8$. It follows that A_3A_3' also lies in a subsystem group D_6 as D_3D_3.

The possibilities for Y are A_3, A_3A_3, A_3A_3', A_5 and D_8 (see 2.8 - if $Y = A_5$ then $V_Y(\lambda_1) \downarrow X = \lambda_2$ and if $Y = D_8$ then $V_Y(\lambda_1) \downarrow X = \lambda_1 + \lambda_3/0$). Note that if $Y = A_5$ then $X < A_5 < D_6$, so $X < A_3A_3'$ also.

The information in Table 8.1 is clear if $Y = A_3$ (restrict from $L(G) \downarrow A_4$). Suppose now that $Y = A_3A_3$. Let A be a fixed copy of A_3, and write elements of Y as $a.b\,(a, b \in A)$. Conjugating X by an element of $N_G(Y)$ interchanging the two A_3 factors, if necessary, we may take $X = \{a.a^\phi : a \in A\}$ for some morphism ϕ of A. Since $N_G(Y)/Y$ contains a subgroup Z_2^2 fixing both factors A_3 and inducing the full group of graph automorphisms, we may replace X by a suitable $N_G(Y)$-conjugate to take $\phi = 1$ or $\phi = \sigma_q$, a Frobenius q-power morphism. Restricting from $L(G) \downarrow A_7$, we obtain

$$
\begin{aligned}
L(G) \downarrow A_3A_3 \;=\; & L(A_3)^2/(\lambda_1 \otimes 0)^2/(0 \otimes \lambda_1)^2/(\lambda_3 \otimes 0)^2/(0 \otimes \lambda_3)^2/ \\
& (\lambda_2 \otimes 0)^2/(0 \otimes \lambda_2)^2/\lambda_1 \otimes \lambda_1/\lambda_3 \otimes \lambda_3/\lambda_1 \otimes \lambda_2/\lambda_2 \otimes \lambda_1/ \\
& \lambda_3 \otimes \lambda_2/\lambda_2 \otimes \lambda_3/\lambda_1 \otimes \lambda_3/\lambda_3 \otimes \lambda_1/0^2.
\end{aligned}
$$

The restrictions $L(G) \downarrow X$ in Table 8.1 follow, using 2.14. When $\phi = 1$, we have $t = 4$, so $C_G(X)^0 = C_{A_7T_1}(X)^0 = A_1T_1$. And when $\phi = \sigma_q$, $t = 2$ and $C_G(X)^0 = C_{A_7T_1}(X)^0 = T_2$.

Next consider $Y = A_3A_3'$. Again $N_G(Y)$ contains an element which interchanges the two simple factors of Y, so as above we may take $X = \{a.a^\phi : a \in A\}$; further, we may assume that $\phi = 1$ or σ_q, since there is an involution centralizing the A_3 and inducing a graph automorphism on the A_3'. Restricting from $L(G) \downarrow A_7'$, we have

$$
\begin{aligned}
L(G) \downarrow A_3A_3' \;=\; & L(A_3)^2/(\lambda_1 \otimes \lambda_3)^2/(\lambda_3 \otimes \lambda_1)^2/(\lambda_2 \otimes 0)^4/(0 \otimes \lambda_2)^4/ \\
& (\lambda_1 \otimes \lambda_1)^2/(\lambda_3 \otimes \lambda_3)^2/\lambda_2 \otimes \lambda_2/0^6.
\end{aligned}
$$

Again $L(G) \downarrow X$ follows, using 2.14. If $\phi = 1$ then as noted before, X also lies in a subsystem group A_5, so $C_G(X)^0 \geq C_G(A_5)^0 = A_2A_1$; from $L(G) \downarrow X$ we have $\dim C_G(X)) \leq 12$, so $C_G(X)^0 = A_2A_1$ or $A_2A_1T_1$, and since $C_{A_5}(X)^0 = 1$, the former must hold. If $\phi = \sigma_q$ then $C_G(X)^0 \geq C_G(D_6)^0 = A_1A_1$; as $L(G) \downarrow X$ has only 6 trivial composition factors, equality holds.

Finally, suppose $Y = D_8$. Here $X < B_7 < Y$, with $V_{B_7}(\lambda_1) \downarrow X = \lambda_1 + \lambda_3$. Since B_7 is centralized by a graph automorphism of D_8, there is just one Y-class of such subgroups X in Y. The restrictions $L(D_8) \downarrow X$ and $V_{D_8}(\lambda_7) \downarrow X$ are given by 2.10 and 2.12, so $L(G) \downarrow X$ is as in Table 8.1, and $C_G(X)^0 = 1$.

Case $X = B_3 \, (p \neq 2)$

The possibilities for Y are D_4, $D_4 D_4$, D_7, A_6, A_7 and A_7'; in the last three cases, X is an irreducible subgroup of Y. When $Y = D_7$, $X < B_3 B_3 < Y$. Note that B_3 has embeddings in D_8 via irreducible representations of degrees 8,8, degrees 1,7,8 or degrees 1,1,7,7. All these embed B_3 in a subsystem subgroup $D_4 D_4$ of D_8. Thus if $Y = A_6$, A_7, A_7' or D_7 then $X < Y < D_8$ and X lies in a subsystem subgroup $D_4 D_4$ of this D_8. So we need consider only $Y = D_4$ or $D_4 D_4$.

If $Y = D_4$ then $N_G(Y)/C_G(Y) \cong S_3$, inducing the full group of graph automorphisms on Y. Therefore there is just one $N_G(Y)$-class of subgroups $X = B_3$ in Y, and $L(G) \downarrow X$ follows by restricting from $L(G) \downarrow Y$. We see that $t = 36$, so $C_G(X)^0 = C_{D_8}(X)^0 = B_4$.

Now let $Y = D_4 D_4$. Both factors D_4 are simply connected, $Z(Y) \cong Z_2^2$ and $N_G(Y)/Y \cong S_3 \times Z_2$ (where the Z_2 interchanges the two simple factors of Y). There are three D_4-classes of subgroups B_3 in D_4, permuted transitively by S_3. Fix a copy D of D_4 and write the elements of Y as $a.b \, (a, b \in D)$; fix also a copy B of B_3 in D. It follows that for each $i \geq 0$, there are just two $N_G(Y)$-classes of diagonal subgroups B_3 in Y, with representatives

$$X_1^{(i)} = \{b.b^{\sigma_p^i} : b \in B\}, \quad X_2^{(i)} = \{b.b^{\tau \sigma_p^i} : b \in B\},$$

where τ is a triality automorphism of D and σ_p is a Frobenius p-power morphism.

Suppose first that $i = 0$, and write $X_j = X_j^{(0)}$ for $j = 1, 2$. Then $Z(X_1) = 1$ and $Z(X_2) \cong Z_2$. By embedding Y in a subsystem group D_8, we see that X_1 lies irreducibly in subsystem groups A_6 and A_7', while X_2 lies in A_7. Restricting from $L(G) \downarrow Y$ (given in 2.1), we have

$$L(G) \downarrow X_1 = (L(D_4)^2 \downarrow X_1)/(\lambda_1 \otimes \lambda_1/\lambda_1^2/0)/\lambda_3 \otimes \lambda_3/\lambda_3 \otimes \lambda_3,$$
$$L(G) \downarrow X_2 = (L(D_4)^2 \downarrow X_2)/(\lambda_1 \otimes \lambda_3/\lambda_3)/\lambda_3 \otimes \lambda_3/(\lambda_3 \otimes \lambda_1/\lambda_3).$$

Hence $L(G) \downarrow X_i \, (i = 1, 2)$ are as in Table 8.1, by 2.14. For X_1 we have $t = 4$, so as $X_1 < A_6$, $C_G(X_1)^0 = C_G(A_6)^0 = A_1 T_1$. Similarly $C_G(X_2)^0 = C_G(A_7)^0 = T_1$.

Now let $i \geq 1$. Embedding Y in D_8, we see that $X_1^{(i)} < D_7$. The restrictions $L(G) \downarrow X_j^{(i)}$ follow using $L(G) \downarrow Y$ again, and clearly $C_G(X_1^{(i)})^0 = T_1$, $C_G(X_2^{(i)})^0 = 1$.

Case $X = C_3 \, (p \neq 2, 3)$

Here Y is A_5 or D_7; in the latter case $V_Y(\lambda_1) \downarrow X = V_X(\lambda_2)$ (see 2.8). When $Y = A_5$, we find $L(G) \downarrow X$ by restricting from $L(G) \downarrow Y$ and using 2.10. Observe that $t = 17$. Since $X < F_4 < G$ here (indeed, X is a Levi subgroup of F_4), $C_G(X)^0$

contains $C_G(F_4)^0 C_{F_4}(X)^0 = G_2 A_1$, so $C_G(X)^0 = G_2 A_1$. Finally, if $Y = D_7$ then $L(G) \downarrow X$ follows from $L(G) \downarrow Y$ using 2.10 and 2.12; as $t = 1$, $C_G(X)^0 = C_G(Y)^0 = T_1$.

Case $X = G_2 \, (p \neq 2, 3, 5, 7)$

The possibilities for Y are D_4, $D_4 D_4$, A_6, E_6 and D_7. In the last case, either $X < B_3 B_3 < Y$ or $V_{D_7}(\lambda_1) \downarrow X = V_X(\lambda_2)$; and if $Y = E_6$ then X is a maximal connected subgroup of Y. When $Y = A_6$, embed Y in D_8 to see that $X < D_4 D_4 < D_8$ also. Thus we need consider only $Y = D_4$, $D_4 D_4$, E_6 or D_7, with X irreducible on $V_{D_7}(\lambda_1)$ in the last case.

If $Y = D_4$ we calculate $L(G) \downarrow X$ from $L(G) \downarrow Y$. We claim that $C_G(X)^0 = F_4$. To see this, let $Z = F_4 < G$. By the F_4 case considered above, $C_G(Z)^0 = G_2$. If T_4 is a maximal torus of Z then $C_G(T_4)'$ has rank at most 4 and contains G_2, so is of type D_4. Therefore X is conjugate to $C_G(Z)^0$, and so $C_G(X)$ contains F_4. Since $t = 52$, it follows that $C_G(X)^0 = F_4$, as claimed.

Now consider $Y = D_4 D_4$. As in previous cases, write the elements of Y as $a.b \, (a, b \in D)$, where $D \cong D_4$. Fix a copy E of G_2 in D. There is just one D_4-class of subgroups G_2 in D_4; hence the representatives of the $N_G(Y)$-classes of diagonal subgroups G_2 in Y are

$$X^{(i)} = \{e.e^{\sigma_p^i} : e \in E\}$$

where as usual, σ_p is a p-power Frobenius morphism and $i \geq 0$. The restriction $L(G) \downarrow X^{(i)}$ follows from $L(G) \downarrow Y$, using 2.10 and 2.14.

Suppose $i = 0$, and write $X = X^{(0)}$. Then $X < A_6 < D_8$. Since this A_6 centralizes a fundamental A_1 in A_8, it follows that $X < E_7$, and by [Se2, p.34-35], $C_{E_7}(X)^0 = A_1$. Hence $C_G(X)^0$ contains $A_1 C_G(E_7) = A_1 A_1$. As $t = 6$, $C_G(X)^0 = A_1 A_1$.

Now suppose $i > 0$. Here $\dim C_G(X^{(i)}) \leq t = 3$. The projection of $X^{(i)}$ to the first D_4 factor of Y centralizes a subgroup F_4 containing the second D_4 factor. Since $C_{F_4}(G_2)^0 = A_1$ (see [Se2, Theorem 1]), it follows that $C_G(X^{(i)})$ contains this A_1, and hence $C_G(X^{(i)})^0 = A_1$.

Next, consider $Y = E_6$. We calculate $L(G) \downarrow X$ from $L(G) \downarrow Y$, using 2.4 and 2.5. As $\dim C_G(X) \leq t = 8$, $C_G(X)^0 = C_G(Y)^0 = A_2$.

Finally, when $Y = D_7$ with $V_Y(\lambda_1) \downarrow X = V_X(\lambda_2)$, $L(G) \downarrow X$ follows from $L(G) \downarrow Y$ using 2.10 and 2.12, and $C_G(X)^0 = C_G(Y)^0 = T_1$.

Case $X = B_2 \, (p \neq 2, 3, 5)$

In this case, Y is A_3, $A_3 A_3$, $A_3 A_3'$, A_4, $A_3 A_4$, $A_4 A_4$, D_5, $A_3 D_5$, D_7, D_8 or E_8. When $Y = D_5$ or $A_3 D_5$, the projection of X to the D_5 factor is either irreducible (with $V_{D_5}(\lambda_1) \downarrow X = V_X(02)$) or lies in $B_2 B_2$ with one of the projections involving a Frobenius twist. If $Y = D_7$ then $V_Y(\lambda_1) \downarrow X = V_X(20)$. If $Y = D_8$ then $V_Y(\lambda_1) \downarrow X = 01 \otimes 01^{(q)}$ for some q. And if $Y = E_8$ then X is a maximal connected subgroup of Y. The subsystems $A_3 A_3$ and $A_3 A_3'$ are as in the case $X = A_3$.

Observe that if $Y = A_3A_3'$ then $X < Y < D_6$, so X lies in a subsystem group D_5 (with $X < B_2B_2 < D_5$). Also, if $Y = A_4$ then $X < Y < D_6$, so $X < A_3A_3'$ as well.

First consider $Y = A_3$ or A_4. Then $L(G) \downarrow X$ is immediate from $L(G) \downarrow Y$ (using 2.10 and 2.14). In the first case ($Y = A_3$), $C_G(X)^0 = C_{D_8}(X)^0 = B_5$; in the second case, $C_G(X)^0 = C_G(Y)^0 = A_4$.

Now assume $Y = A_3A_3$ or A_3A_3'. Then X lies in a subgroup $X_0 \cong A_3$ which is a diagonal subgroup of Y, and $L(G) \downarrow X_0$ is given in Table 8.1. Restricting to X using 2.14, we obtain $L(G) \downarrow X$. When $Y = A_3A_3'$, $C_G(X)^0$ is clearly as in Table 8.1. For $Y = A_3A_3$, observe that D_8 contains tensor product subgroups $C_2 \otimes C_2$ (with a factor C_2 lying in Y) and $(C_2 \otimes C_1) \oplus (C_2^{(q)} \otimes C_1)$, exhibiting the centralizers given in the table.

Next let $Y = A_3A_4$ or A_4A_4. Then $L(G) \downarrow X$ follows from $L(G) \downarrow Y$ (given in 2.1), using 2.10 and 2.14. When $Y = A_3A_4$ and no Frobenius twist is involved in the embedding of X in Y, we have $X = B_2 < B_2 \otimes B_1 < B_7 < D_8$, so $C_G(X)^0 = B_1$. In other cases, $C_G(X)^0$ is clearly as in the table. Note also that if $Y = A_4A_4$, then X lies in a subgroup B_2B_2 of Y which is centralized by an involution $t \in N_G(Y)$ acting as a graph automorphism on both factors A_4. Since $C_G(X)^0 = 1$, we have $X < B_2B_2 < C_G(t) = D_8$, and $V_{D_8}(\lambda_1) \downarrow B_2B_2 = 01 \otimes 01$. If no Frobenius twist is involved in the embedding of X in Y, then $V_{D_8}(\lambda_1) \downarrow X = 02/10/00$, so $X < A_3D_5 < D_8$; and if a Frobenius q-twist is involved, then $V_{D_8}(\lambda_1) \downarrow X = 01 \otimes 01^{(q)}$.

Consider now $Y = D_8$, with $V_Y(\lambda_1) \downarrow X = 01 \otimes 01^{(q)}$. There are two Y-classes of such subgroups X. One class contains subgroups lying in A_4A_4, as shown in the previous paragraph. For the other class, $L(G) \downarrow X$ follows from $L(G) \downarrow Y$ using 2.10 and 2.12, and $C_G(X)^0 = 1$.

Now suppose $Y = D_5$. If $X < B_2B_2 < Y$ then X lies in A_3A_3', a case considered above. So assume $V_Y(\lambda_1) \downarrow X = V_X(02)$. Then $L(G) \downarrow X$ follows from $L(G) \downarrow Y$, using 2.10 and 2.12, and $C_G(X)^0 = C_Y(X)^0 = A_3$. Note that there is just one class of such subgroups X in $N_G(Y) = (A_3D_5).2$.

Next consider $Y = A_3D_5$. Restricting from $L(G) \downarrow D_8$ given in 2.1, we have

$$L(G) \downarrow A_3D_5 = L(A_3)/L(D_5)/\lambda_2 \otimes \lambda_1/\lambda_1 \otimes \lambda_4/\lambda_3 \otimes \lambda_5$$

(up to a graph automorphism). Suppose that the projection of X in the D_5 factor is irreducible on $V_{D_5}(\lambda_1)$. Then for some $q, q' \geq 1$, we have, using 2.10 and 2.12,

$$L(G) \downarrow X = (L(A_3) \downarrow X)/(12/02)^{(q')}/(10/00)^{(q)} \otimes 02^{(q')}/01^{(q)} \otimes 11^{(q')}/01^{(q)} \otimes 11^{(q')}.$$

Using 2.14, we see that $L(G) \downarrow X$ is as in Table 8.1. Clearly $C_G(X)^0 = 1$.

Now suppose that the projection of X in D_5 is reducible, so $X < B_2B_2B_2 < D_8$ and $N_{D_8}(B_2B_2B_2)$ acts as S_3 on the factors. Fixing a copy B of B_2, and writing elements of this $B_2B_2B_2$ as $a.b.c\,(a, b, c \in B)$, we may assume

$$X = \{b.b^\phi.b^\psi : b \in B\}$$

where ϕ, ψ are morphisms of B. Replacing X by a suitable D_8-conjugate, we can take $\phi = \sigma_p^i, \psi = \sigma_p^j$ for some i, j and some Frobenius p-power morphism σ_p. If i or j is 0, or if $i = j$, then $X < A_3 A_4 < D_8$, a case considered above, so we can assume that $p^i = q, p^j = q'$ and $1, q, q'$ are distinct. For each q, q' there is just one D_8-class of such subgroups X, and $L(G) \downarrow X$ is easily deduced from $L(G) \downarrow A_3 D_5$ given above.

Now let $Y = D_7$, with $V_Y(\lambda_1) \downarrow X = V_X(20)$. There is just one class of such subgroups X in $N_G(Y) = (D_7 T_1).2$, and $L(G) \downarrow X$ follows from $L(G) \downarrow Y$ and 2.10, 2.12. Clearly $C_G(X)^0 = C_G(Y)^0 = T_1$.

Finally, when $Y = E_8$ and X is a maximal connected subgroup of Y, $L(G) \downarrow X$ is given in 2.4.

Case $X = A_2 \, (p \neq 2, 3, 5)$

Here Y is $A_2^k \, (k \leq 4)$, A_5, $A_2 A_5$, D_4, $A_2 D_4$, $D_4 D_4$, A_7, A_7', A_8, E_6, $A_2 E_6$ or E_7. When Y has a factor A_5, D_4, A_7 or A_7', the usual module for this factor restricts irreducibly to X as $V_X(20)$ or $V_X(11)$ (possibly twisted); when $Y = A_8$, $V_Y(\lambda_1) \downarrow X$ is $10 \otimes 10^{(q)}$ or $10 \otimes 01^{(q)}$; and when Y has a factor E_6 or E_7, the projection of X is maximal in this factor.

If $Y = A_7$ or A_7' then $X < Y < D_8$, so $X < D_4 D_4$ also. Now suppose $Y = D_4$. Then $X < Y.S_3 < F_4 < G$. The group $Y.S_3$ contains an element t of order 3 such that $C_Y(t)^0 = A_2$, of adjoint type (see [GL, 9.1]). Replacing X by an $N_G(Y)$-conjugate, we can take $X = C_Y(t)^0$. Observe that $C_{F_4}(t)^0 = A_2 A_2$, since it is not B_3 or C_3 (as neither of these contains an adjoint A_2). This F_4-subsystem $A_2 A_2$ lies in a subsystem group A_2^3 in E_6. Thus when $Y = D_4$, X lies in A_2^3, essentially embedded. Similarly, if $Y = A_2 D_4$ then X lies in A_2^4 as well.

First consider $Y = A_2^4$. The restriction $L(G) \downarrow Y$ is given in 2.2. Fixing a copy A of A_2, and writing elements of Y as $a_1.a_2.a_3.a_4 \, (a_i \in A)$, we may take

$$X = \{a.a^{\phi_1}.a^{\phi_2}.a^{\phi_3} : a \in A\},$$

where the ϕ_i are morphisms of A. Write X_1 for the subgroup $\{a.1.1.1 : a \in A\}$ of Y, and define X_2, X_3, X_4 similarly, so that $Y = X_1 X_2 X_3 X_4$. By [Ca], $N_G(Y)/Y \cong GL_2(3)$, with central element -1 acting as a graph automorphism on each X_i, and factor group $GL_2(3)/\langle -1 \rangle \cong S_4$ permuting the X_i naturally. Write S for this group S_4.

Suppose now that none of the ϕ_i involves a Frobenius twist. If τ is a graph automorphism of A of order 2, then we can take each of the ϕ_i to be either 1 or τ, giving 8 possibilities for the group X. We claim that S permutes the corresponding 8 Y-classes transitively, so that there is just one $N_G(Y)$-class of such subgroups X of Y. To see this, consider $X_2 X_3 X_4$ as a subgroup of $C_G(X_1) = E_6$. With the labelling of the extended E_6-diagram as in the Introduction, take X_2, X_3, X_4 to correspond to the subsystems $\langle \alpha_1, \alpha_3 \rangle$, $\langle \alpha_6, \alpha_5 \rangle$, $\langle -\alpha_0, \alpha_2 \rangle$, respectively. By [DL, p.50], $N_{E_6}(X_2 X_3 X_4)/X_2 X_3 X_4$ is a subgroup S_3 of S, generated by δ_1 and δ_2, where

$$\delta_1 = (\alpha_1 \, \alpha_6 - \alpha_0)(\alpha_3 \, \alpha_5 \, \alpha_2), \quad \delta_2 = (\alpha_1 \, \alpha_2)(\alpha_3 - \alpha_0)(\alpha_5 \, \alpha_6).$$

Therefore we can choose a copy X of A_2 in Y which is fixed by δ_1 but not by δ_2. The stabilizer in S of this group X must be just $\langle \delta_1 \rangle \cong Z_3$, so the transitivity of S on the above 8 groups follows. Now the restriction $L(G) \downarrow X$ in this case follows from 2.2 and 2.14. We find that $t = 2$, so $C_G(X)^0 = C_G(A_2D_4)^0 = C_{E_6}(D_4)^0 = T_2$ (as noted above, X also lies in A_2D_4).

Next, suppose that just one of the ϕ_i involves a nontrivial Frobenius morphism σ_q, say ϕ_3. Then we can take ϕ_1, ϕ_2 to be 1 or τ, and ϕ_3 to be σ_q or $\tau\sigma_q$, again giving 8 possibilities for the group X. The subgroup of S stabilizing the corresponding set of 8 Y-classes is S_3 (fixing X_4), and it has two orbits (since, by the above, the stabilizer of a class lies in Z_3). Thus there are two $N_G(Y)$-classes of such subgroups X of Y. For both classes, $L(G) \downarrow X$ follows from 2.2 and 2.14. One class contains subgroups lying in A_2D_4, and $C_G(X)^0 = C_G(A_2D_4)^0 = T_2$ for this class; for the other, $C_G(X)^0 = 1$. A similar analysis applies when each ϕ_i is σ_q or $\tau\sigma_q$.

Now suppose that $\phi_1 = 1$ or τ and ϕ_2, ϕ_3 are σ_q or $\tau\sigma_q$ (the same q for both). The subgroup of S stabilizing the corresponding set of 8 Y-classes is $S_2 \times S_2$ acting semiregularly, hence with two orbits. So there are two $N_G(Y)$-classes of subgroups X. As usual, $L(G) \downarrow X$ follows from 2.2 and 2.14, and $C_G(X)^0 = 1$ for both classes.

Assume next that $\phi_1 = 1$ or τ, $\phi_2 = \sigma_q$ or $\tau\sigma_q$, and $\phi_3 = \sigma_{q'}$ or $\tau\sigma_{q'}$ (where as always, $1, q, q'$ are distinct). This time, only S_2 acts on the 8 Y-classes, with four orbits, so there are four classes of subgroups X. Similarly, there are four classes if $\phi_1, \phi_2 = \sigma_q$ or $\tau\sigma_q$ and $\phi_3 = \sigma_{q'}$ or $\tau\sigma_{q'}$.

Lastly, take $\phi_i = \sigma_{q_i}$ or $\tau\sigma_{q_i}$, where $1, q_1, q_2, q_3$ are distinct. There are 8 classes of subgroups X in this case, and $L(G) \downarrow X$ follows from 2.2. This completes the case where $Y = A_2^4$.

Now consider $Y = A_2^3$. With the above notation, we can take

$$X = \{a.a^{\phi_1}.a^{\phi_2} : a \in A\}.$$

Also $N_G(Y)/C_G(Y) \cong S_3$. If ϕ_1, ϕ_2 are 1 or τ, then from the above analysis, we see that this S_3 has two orbits on the 4 Y-classes containing such subgroups X, giving two $N_G(Y)$-classes of subgroups X. The restrictions $L(G) \downarrow X$ follow from 2.2 and 2.14. Using 2.3, we see that a subsystem group A_2A_2 in a subgroup F_4 of G contains representatives of both classes of subgroups X. For one class, the group X lies in D_4, and $C_G(X)^0 = C_G(D_4)^0 = D_4$; for the other, $C_G(X)^0 = C_G(F_4)^0 = G_2$.

When $\phi_1 = 1$ or τ and $\phi_2 = \sigma_q$ or $\tau\sigma_q$, there are three $N_G(Y)$-classes of subgroups X; two of these classes contain groups lying in F_4 and having centralizer G_2, and the other class contains groups with centralizer $C_G(A_2^3) = A_2$. Similar calculations yield Table 8.1 when ϕ_1, ϕ_2 are both σ_q or $\tau\sigma_q$. Finally, if $\phi_1 = \sigma_q$ or $\tau\sigma_q$ and $\phi_2 = \sigma_{q'}$ or $\tau\sigma_{q'}$, there are four $N_G(Y)$-classes of subgroups X, all with centralizer A_2.

Next, let $Y = A_2^2$, so $X = \{a.a^\phi : a \in A\}$. If ϕ is 1 or τ then there is one $N_G(Y)$-class of subgroups X; this group X lies in F_4 and is invariant under a maximal torus of F_4, so $C_{F_4}(X)^0 = A_2$. Therefore $C_G(X)^0 = C_{F_4}(X)^0 C_G(F_4)^0 = A_2G_2$. And if $\phi = \sigma_q$ or $\tau \sigma_q$ then there is one $N_G(Y)$-class of subgroups X, with $C_G(X)^0 = C_G(Y)^0 = A_2 A_2$.

If $Y = A_2$ then the conclusions in Table 8.1 are clear.

Now let $Y = A_5$. Then Y contains one conjugacy class of irreducible subgroups $X = A_2$, and $L(G) \downarrow X$ follows from $L(G) \downarrow Y$, using 2.10 and 2.14. Also $C_G(X)^0 = C_G(Y)^0 = A_1 A_2$.

Consider next $Y = A_2 A_5$. Here $N_G(Y)/C_G(Y) = \langle w \rangle \cong Z_2$, where w acts as a graph automorphism on both factors of Y. If the projections of X to the factors of Y involve no Frobenius twist, then there are two $N_G(Y)$-classes of such subgroups X. To calculate $L(G) \downarrow X$, we first restrict from $L(G) \downarrow A_2 E_6$ (given in 2.1), using 2.3, to obtain

$$L(G) \downarrow A_2 A_5 = L(A_2)/L(A_5)/(0 \otimes \lambda_3)^2/(\lambda_1 \otimes \lambda_5)^2/\lambda_1 \otimes \lambda_2/(\lambda_2 \otimes \lambda_1)^2/\lambda_2 \otimes \lambda_4/0^3.$$

From this we find $L(G) \downarrow X$, using 2.10 and 2.14. For both classes, $C_G(X)^0 = C_G(Y)^0 = A_1$. (We shall see below that one of these classes contains subgroups which also lie in a subsystem group $D_4 D_4$.) Similarly, when one of the projections involves a Frobenius twist, there are two $N_G(Y)$-classes of subgroups, both with $C_G(X)^0 = A_1$.

It remains to handle the cases where

$$Y = D_4 D_4, A_8, E_6, A_2 E_6 \text{ or } E_7.$$

Let $Y = D_4 D_4$. We have already observed that $N_G(Y)/Y \cong S_3 \times Z_2$, and that an irreducible copy of A_2 in D_4 is centralized by an outer automorphism of order 3. Let D be a fixed copy of D_4, and A a fixed irreducible copy of A_2 in D, and write elements of Y as $a.b$ $(a, b \in D)$. Then for each $i \geq 0$, there are two $N_G(Y)$-classes of diagonal subgroups A_2 in Y, with representatives

$$X_1^{(i)} = \{a.a^{\sigma_p^i} : a \in A\}, \quad X_2^{(i)} = \{a.a^{\tau \sigma_p^i} : a \in A\}$$

(where as usual, τ is a graph automorphism of A). We obtain $L(G) \downarrow X_j^{(i)}$ $(j = 1, 2)$ by restricting from $L(G) \downarrow Y$ (given in 2.1) and using 2.14. Observe that for any given i, the restrictions $L(G) \downarrow X_1^{(i)}$ and $L(G) \downarrow X_2^{(i)}$ have the same composition factors: these subgroups provide the only exceptions in the statement of Theorem 4 (we shall demonstrate below that $X_1^{(i)}$ is not conjugate to $X_2^{(i)}$). If $i > 0$ then $C_G(X_j^{(i)})^0 = 1$. For $i = 0$, write $X_j = X_j^{(0)}$ $(j = 1, 2)$. From $L(G) \downarrow X_j$ we see that $\dim C_G(X_j) \leq t = 3$. By embedding Y in a subsystem group D_8, we can take $X_1 < A_7 < D_8$ and $X_2 < A_7' < D_8$ (as τ is induced by a graph automorphism of D_8). Then $C_G(X_2)^0 = C_G(A_7')^0 = A_1 \cong SL_2$ while $C_G(X_1)^0 = A_1$ by [Se2, p.46]. Moreover, X_1 and X_2 are not G-conjugate, since maximal tori of $C_G(X_1)$, $C_G(X_2)$ have different centralizers, namely E_7, A_7, respectively. To see that $X_1^{(i)}$ and $X_2^{(i)}$ are not conjugate for $i > 0$, suppose to the contrary that $(X_1^{(i)})^g = X_2^{(i)}$ for some

$g \in G$. From the restriction of $L(G) \downarrow X_j^{(i)}$ (given in Table 8.1), we see that the restrictions of $L(G)/L(Y)$ and $L(Y)$ have no common composition factors. It follows that g stabilizes $L(Y) = L(D_4) \oplus L(D_4)$. Therefore g lies in the maximal subgroup $N_G(D_4 D_4) = N_G(Y)$. But we know that $X_1^{(i)}$ and $X_2^{(i)}$ are not conjugate in $N_G(Y)$.

Next, observe that there is a 3-element $x \in N_G(Y)$ such that $C_Y(x) = A_2 A_2$ (adjoint groups) containing $X_1^{(i)}$ and $X_2^{(i)}$. The element x cycles the three factors $\lambda_i \otimes \lambda_j$ in $L(G) \downarrow D_4 D_4$, and acts on each $L(D_4)$ with fixed space of dimension 8. Hence $\dim C_{L(G)}(x) = 80$. It follows that $C_G(x) = A_8$, with $V_{A_8}(\lambda_1) \downarrow A_2 A_2 = 10 \otimes 10$. When $i > 0$, this accounts for the case $Y = A_8$. And when $i = 0$, X_1 lies in a subsystem group $A_2 A_5$ in this A_8.

When $Y = E_6$, X is a maximal connected subgroup of Y, and $L(G) \downarrow X$ follows from $L(G) \downarrow Y$, using 2.4 and 2.5.

Now suppose $Y = A_2 E_6$ and the projection of X in the E_6 factor is maximal connected. By [Ca], $N_G(Y)/Y = \langle w \rangle \cong Z_2$, where w induce a graph automorphism on both factors A_2 and E_6. Fix a copy A of A_2, maximal connected in E_6. By [Te2, p.314], E_6 contains $A\langle \tau \rangle$, where τ is a graph automorphism of A of order 2. Hence E_6 contains two classes of maximal connected subgroups A_2, interchanged by w. If neither projection of X to the factors of Y involves a Frobenius twist, it follows that there is just one $N_G(Y)$-class of such subgroups X; similarly, if one of the projections involves a Frobenius twist, there is also just one class. In any case, $C_G(X)^0 = 1$.

Finally, when $Y = E_7$ with X maximal connected in Y, the restriction $L(G) \downarrow X$ follows from $L(G) \downarrow Y$ and 2.4, 2.5, and $C_G(X)^0 = C_G(Y)^0 = A_1$.

This completes the proof of all the information in Table 8.1, hence finishing the proof of Theorem 5 for $G = E_8$.

We conclude the proof of Theorem 5 for the remaining types of exceptional group G, by showing how to deduce the information in Tables 8.2-8.5 from that in Table 8.1.

Suppose that X is a simple closed connected subgroup of E_8, with $\mathrm{rank}(X) \geq 2$ and $p = 0$ or $p > N(X, E_8)$, so that X is as in Table 8.1. Regard E_7 as $C_{E_8}(\bar{A}_1)$, where $\bar{A}_1$ is a fundamental subgroup SL_2. Thus X lies in E_7 if and only if $C_{E_8}(X)$ contains $\bar{A}_1$, and so we can read off from the group $C_{E_8}(X)^0$ given in Table 8.1 whether or not X lies in E_7.

Suppose we have two conjugates X, X^g $(g \in E_8)$ lying in E_7. Then $\bar{A}_1, \bar{A}_1^{g^{-1}} \leq C_{E_8}(X)$. If $\bar{A}_1$ and $\bar{A}_1^{g^{-1}}$ are conjugate in $N_{E_8}(X)$, say $\bar{A}_1^{g^{-1}} = \bar{A}_1^n$ with $n \in N_{E_8}(X)$, then $ng \in N_{E_8}(\bar{A}_1) = \bar{A}_1 E_7$, and $X^{ng} = X^g$, so X and X^g are E_7-conjugate. On the other hand, if $\bar{A}_1$ and $\bar{A}_1^{g^{-1}}$ are not $N_{E_8}(X)$-conjugate then X, X^g are not E_7-conjugate: for if $X^g = X^e$ with $e \in E_7$, then $eg^{-1} \in N_{E_8}(X)$ and $\bar{A}_1^{eg^{-1}} = \bar{A}_1^{g^{-1}}$, which means that $\bar{A}_1$ and $\bar{A}_1^{g^{-1}}$ are $N_{E_8}(X)$-conjugate. We have now shown

1. $X \leq E_7$ if and only if $C_{E_8}(X)$ contains $\bar{A}_1$, and

2. for $X \leq E_7$, the E_7-orbits on $\{X^g \leq E_7 : g \in E_8\}$ correspond to the $N_{E_8}(X)$-classes of subgroups $\bar{A}_1$ in $C_{E_8}(X)$.

In order to use (2), we assert that $N_{E_8}(X)$ fixes every factor in $C_{E_8}(X)$ containing a conjugate of $\bar{A}_1$: this is clear from Table 8.1, except for the cases $X = A_3, C_G(X)^0 = \bar{A}_1\bar{A}_1$ and $X = A_2, C_G(X)^0 = \bar{A}_2\bar{A}_2$. In the first case, $X < A_3A_3'^q < D_6$ and $N_G(X) \leq N_G(\bar{A}_1\bar{A}_1) = N_G(Y)$ where Y is a subsystem subgroup $A_1A_1D_6$. Here $N_G(Y)/Y \cong Z_2$, with nontrivial element interchanging the A_1 factors and inducing an element in the coset of a graph automorphism of the D_6 factor. Since X is not normalized by such an element, the assertion follows for this case. The other case is similar, using the fact that $N_G(X) < N_G(A_2^4)$.

It is now straightforward to deduce the information on conjugacy classes and centralizers in Table 8.2. The restrictions $L(E_7) \downarrow X$ can also be deduced by referring to the proofs for Table 8.1.

Notice that for $X = B_2$ or C_3, the restrictions on p for E_7 and E_8 are different ($p = 5$ is allowed for $(X, G) = (B_2, E_7)$, and $p = 3$ for $(X, G) = (C_3, E_7)$). However this makes no difference to the above calculations, as the restrictions on p are used only in proving Theorem 1, and not in the calculations.

For $G = E_6$, we deduce Table 8.3 using the same reasoning as for E_7, replacing $\bar{A}_1$ by $\bar{A}_2$ ($= C_{E_8}(G)$). For $G = F_4$, we deduce Table 8.4 by a similar (but very much shorter) analysis to that for $G = E_8$ (note that for this case we must allow subsystems of types B_2, B_3, C_3). For Table 8.5 there is nothing to prove.

All parts of Theorem 5 are now established.

6 Labelled diagrams and conjugacy: proof of Theorems 4 and 6

Let G be an exceptional simple algebraic group over an algebraically closed field of characteristic p. For the purposes of proving Theorems 4 and 6, we may assume that G is of adjoint type.

We begin this section by proving Theorem 4 for simple subgroups of rank at least 2. Suppose then that X_1 and X_2 are closed connected simple subgroups of G of the same type, and of rank at least 2. Assume also that $p = 0$ or $p > N(X_1, G)$, and that X_1 and X_2 have the same composition factors on $L(G)$ (counting multiplicities). By Theorem 5, X_1 and X_2 are as in Tables 8.1-8.5. It is immediate from the restrictions $L(G) \downarrow X_i$ given in these tables that either X_1 and X_2 are conjugate in Aut G, or $X_1 \cong X_2 \cong A_2$, $G = E_8$ and X_1, X_2 lie in a subsystem subgroup $D_4 D_4$ of G. Hence Theorem 4 holds when X_1 has rank at least 2.

The rest of the section is devoted to the proof of Theorems 4 and 6 for subgroups of type A_1. Using the idea of [Se2, §2] (which goes back to Dynkin [Dy]), we associate with any subgroup $X \cong A_1$ of G a *labelled diagram* in the following way. Let $T_X = \{T(c) : c \in K^*\}$ be a maximal torus of X (where $T(c)$ corresponds to the matrix $\mathrm{diag}\,(c, c^{-1})$ in SL_2), and choose a maximal torus T of G containing T_X. Let $\Sigma = \Sigma(G)$ be the root system of G relative to T, and for $\beta \in \Sigma(G)$ let e_β be a weight vector for T corresponding to β. Then there exist integers l_β ($\beta \in \Sigma(G)$) such that for all $c \in K^*$,

$$T(c)e_\beta = c^{l_\beta} e_\beta.$$

We can choose a fundamental system $\Pi = \Pi(G)$ such that $l_\alpha \geq 0$ for all $\alpha \in \Pi$ (this just amounts to choosing an appropriate fundamental region).

Definition   The *labelled diagram* for the subgroup $X \cong A_1$ is defined to be the Dynkin diagram of G, with each node $\alpha \in \Pi$ labelled by the non-negative integer l_α.

Proposition 6.2 below shows that the labelled diagram of X is independent (up to graph automorphisms) of the choice of fundamental system Π.

Let l be the rank of G, and let $W = W(G)$ be the Weyl group of G, operating on the Euclidean space $\mathbf{R}^l = \mathbf{R} \otimes \Pi$ with the usual inner product $(\ ,\ )$. Write Σ^+, Σ^- for the positive and negative roots in Σ with respect to Π.

Lemma 6.1 *Suppose that $v \in \mathbf{R}^l$ and Π, Π' are fundamental systems in $\Sigma(G)$ such that $(v, \alpha) \geq 0$ for all $\alpha \in \Pi \cup \Pi'$. Then there exists $w \in W(G)$ such that $w(\Pi) = \Pi'$ and $w(v) = v$.*

Proof. Certainly there exists $w \in W(G)$ with $w(\Pi) = \Pi'$. We argue by induction on $l(w)$ that $w(v) = v$. This is clear if $w = 1$. Otherwise, we can choose $\alpha \in \Pi$ such that $w(\alpha) \in \Sigma^-$. Since $w(\alpha) \in \Pi'$, we have $(v, w(\alpha)) \geq 0$. On the other hand,

$(v, w(\alpha)) \leq 0$ as $w(\alpha) \in \Sigma^-$. Therefore $(v, w(\alpha)) = 0$. Hence $(w^{-1}(v), \alpha) = 0$ and so $s_\alpha w^{-1}(v) = w^{-1}(v)$, where s_α is the reflection corresponding to α.

Let $\Pi'' = ws_\alpha(\Pi)$, another fundamental system. Observe that for all $\beta \in \Pi$,

$$(v, ws_\alpha(\beta)) = (s_\alpha w^{-1}(v), \beta) = (w^{-1}(v), \beta) = (v, w(\beta)) \geq 0$$

(as $w(\beta) \in \Pi'$). As $l(ws_\alpha) < l(w)$, this implies by induction that $ws_\alpha(v) = v$. Consequently

$$v = s_\alpha w^{-1}(v) = w^{-1}(v)$$

and hence $w(v) = v$. $\square$

Proposition 6.2 *Suppose that Π, Π' are fundamental systems such that $l_\alpha \geq 0$ for all $\alpha \in \Pi \cup \Pi'$ (where l_α is as defined above). Then there exists $w \in W(G)$ such that $w(\Pi) = \Pi'$ and $l_{w(\alpha)} = l_\alpha$ for all $\alpha \in \Pi$. (Thus the labelled diagram of $X \cong A_1$ is uniquely determined up to graph automorphisms.)*

Proof. Choose $v \in \mathbf{R}^l$ such that $(v, \beta) = l_\beta$ for all $\beta \in \Sigma(G)$. By 7.1, there exists w such that $w(\Pi) = \Pi'$ and $w(v) = v$. Then for $\alpha \in \Pi$,

$$l_{w(\alpha)} = (v, w(\alpha)) = (w^{-1}(v), \alpha) = (v, \alpha) = l_\alpha. \quad \square$$

We now state the two main results of this section, which together give Theorems 4 and 6 for subgroups of type A_1.

Theorem 6.3 *Suppose that X_1, X_2 are subgroups of G isomorphic to A_1 and having the same composition factors on $L(G)$ (counting multiplicities), and that $p = 0$ or $p > N(A_1, G)$. Then X_1 and X_2 have the same labelled diagram (up to graph automorphisms of G).*

Theorem 6.4 *Suppose that X_1, X_2 are subgroups of G isomorphic to A_1 and having the same labelled diagram, and that $p = 0$ or $p > N(A_1, G)$. Then X_1 and X_2 are conjugate in $\mathrm{Aut}\, G$.*

Proof of Theorem 6.3

We thank Dr Ross Lawther for providing the following shortened version of our original proof of this result.

Suppose that X_1 and X_2 are subgroups A_1 of G having the same composition factors on $L(G)$. Take X_1, X_2 to have labelled diagrams with labels m_i, n_i, respectively, on the ith node of the Dynkin diagram of G. Since X_1, X_2 have the same composition factors, the sets of values taken over all positive roots are the same for both; in other words, the collection of positive integers $\sum a_i m_i$ for positive roots $\sum a_i \alpha_i$ is the same as the collection $\sum a_i n_i$ (counting multiplicities). We take Dynkin diagrams to be labelled as in the Introduction, and denote a root $\sum a_i \alpha_i$ by $a_1 \ldots a_l$.

Case $G = E_6$ The highest roots are

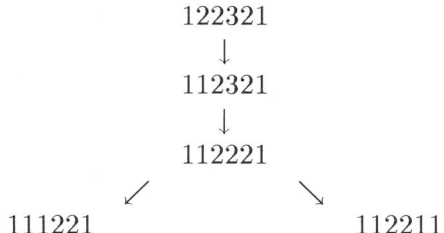

Thus we can identify the values taken on the three highest roots, and so by subtraction we have $m_2 = n_2$ and $m_4 = n_4$; also by applying a graph automorphism, we may assume $m_3 = n_3$. By subtracting these values from that of 122321 we have $m_1 + 2m_5 + m_6 = n_1 + 2n_5 + n_6$. By summing all positive roots we have

$$16m_1 + 22m_2 + 30m_3 + 42m_4 + 30m_5 + 16m_6 = 16n_1 + 22n_2 + 30n_3 + 42n_4 + 30n_5 + 16n_6,$$

so using the equations above we obtain $m_5 = n_5$ and $m_1 + m_6 = n_1 + n_6$. By comparing the least values taken outside $\langle \alpha_2, \alpha_3, \alpha_4, \alpha_5 \rangle$ (where we know the values are identical), we see that either m_1 or m_6 must equal either n_1 or n_6: so either $m_1 = n_1$, $m_6 = n_6$ and the labellings are the same, or $m_1 = n_6$, $m_6 = n_1$. In the latter case the least value outside $\langle \alpha_2, \alpha_3, \alpha_4, \alpha_5 \rangle \cup \{\alpha_1, \alpha_6\}$ must occur at either $\alpha_1 + \alpha_3$ or $\alpha_5 + \alpha_6$: thus either $n_1 + n_3$ or $n_5 + n_6$ must equal either $n_6 + n_3$ or $n_1 + n_5$ – so either $n_1 = n_6$ or $n_3 = n_5$, and the labellings are the same (up to graph automorphism).

Case $G = E_8$ Here the highest roots are

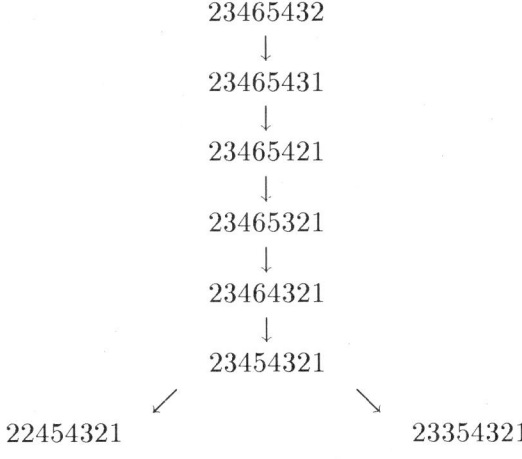

Thus we can identify the values taken on the six highest roots, and so by subtraction we have $m_i = n_i$ for $4 \leq i \leq 8$, and $2m_1 + 3m_2 + 4m_3 = 2n_1 + 3n_2 + 4n_3$. By summing all positive roots we have

$$92m_1 + 136m_2 + 182m_3 + 270m_4 + 220m_5 + 168m_6 + 114m_7 + 58m_8$$

$$= 92n_1 + 136n_2 + 182n_3 + 270n_4 + 220n_5 + 168n_6 + 114n_7 + 58n_8,$$

so by using the equations above we have $m_2 + m_3 = n_2 + n_3$ and $2m_1 + m_3 = 2n_1 + n_3$. If the next highest value occurs at the same root for the two labellings, by subtraction we either have $m_2 = n_2$ or $m_3 = n_3$, and the labellings are the same; if not, we must have $m_2 = n_3$, $m_3 = n_2$ and so $m_1 = n_1 + \frac{1}{2}(n_3 - n_2)$. Since the values taken on $\langle \alpha_i : 2 \leq i \leq 8 \rangle$ are then the same, and the least value taken outside these roots occurs at α_1, we must have $m_1 = n_1$, so $n_3 = n_2$ and the labellings are the same.

Case $G = E_7$ The highest roots are

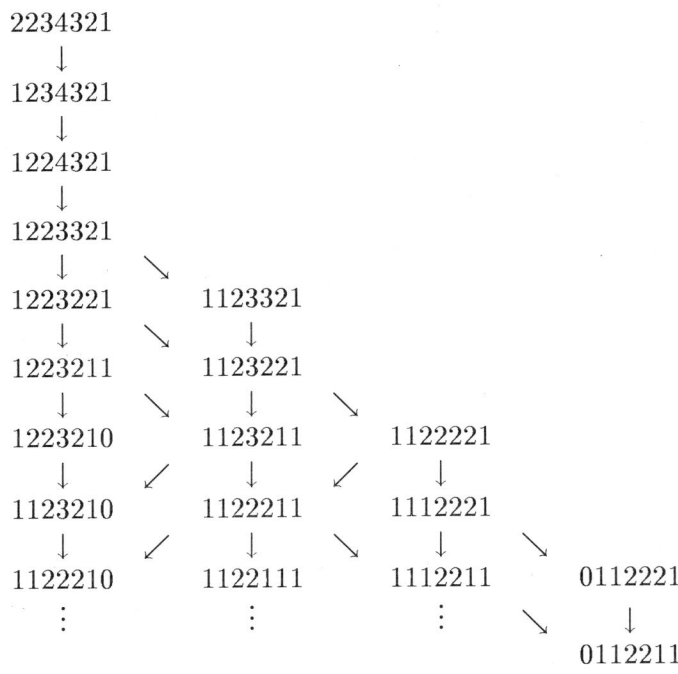

(we shall not need all the roots of height 8, but the fact that all are less than 1223211 will be important). Thus we can identify the values taken on the four highest roots, and so by subtraction we have $m_1 = n_1$, $m_3 = n_3$, $m_4 = n_4$, and $2m_2 + 3m_5 + 2m_6 + m_7 = 2n_2 + 3n_5 + 2n_6 + n_7$. By summing all positive roots we have

$$34m_1 + 49m_2 + 66m_3 + 96m_4 + 75m_5 + 52m_6 + 27m_7 =$$

$$34n_1 + 49n_2 + 66n_3 + 96n_4 + 75n_5 + 52n_6 + 27n_7,$$

so by using the equations above we have $2m_7 + 2m_6 - m_2 = 2n_7 + 2n_6 - n_2$. Let $\alpha = 1223321$. Then there are three possibilities for the roots where the next two highest values occur for the labelling $\{n_i\}$ (and similarly for $\{m_i\}$):

(i) $1223221 = \alpha - \alpha_5$, $1223211 = \alpha - \alpha_5 - \alpha_6$ if $n_5 + n_6 < n_2$;

(ii) $1223221 = \alpha - \alpha_5$, $1123321 = \alpha - \alpha_2$ if $n_5 \leq n_2 \leq n_5 + n_6$;

(iii) $1123321 = \alpha - \alpha_2$, $1223221 = \alpha - \alpha_5$ if $n_2 < n_5$.

If (i) holds for both $\{n_i\}$ and $\{m_i\}$, then by subtraction we have $m_5 = n_5$ and $m_6 = n_6$. Then the above equations imply that the remaining values agree, and the labellings are the same. If (ii) holds for both, or if (iii) does, we have $m_5 = n_5$ and $m_2 = n_2$, and again the labellings agree. Thus we need only consider the cases where different possibilities hold for $\{n_i\}$ and $\{m_i\}$: by symmetry we need only treat three cases.

Case A: (i) holds for $\{n_i\}$ and (ii) for $\{m_i\}$, so that $n_5 + n_6 < n_2$ and $m_5 \leq m_2 \leq m_5 + m_6$. We have $m_5 = n_5$ and $m_2 = n_5 + n_6$: substituting in the equations above gives $m_7 + 2m_6 = n_7 - 2n_5 + 2n_2$, $2m_7 + 2m_6 = 2n_7 + 3n_6 + n_5 - n_2$, so $m_7 = n_7 + 3n_6 + 3n_5 - 3n_2$, $m_6 = -\frac{3}{2}n_6 - \frac{5}{2}n_5 + \frac{5}{2}n_2$. The seventh highest value for $\{n_i\}$ must occur at either 1123321 or 1223210 – the difference is $\alpha_7 + \alpha_6 + \alpha_5 - \alpha_2$, and as $n_7 + n_6 + n_5 - n_2 \geq \frac{1}{3}n_7 + n_6 + n_5 - n_2 = \frac{1}{3}m_7 \geq 0$ it must be the former. The corresponding root for $\{m_i\}$ must be 1223211 or 1123221 – if it is the former, by subtraction we obtain $n_2 = m_5 + m_6 = -\frac{3}{2}n_6 - \frac{3}{2}n_5 + \frac{5}{2}n_2$, so $\frac{3}{2}n_2 = \frac{3}{2}n_6 + \frac{3}{2}n_5$, contrary to $n_5 + n_6 < n_2$. Thus it must be the latter, and we have $n_2 = m_5 + m_2 = n_6 + 2n_5$. (Note that $n_5 > 0$, as $n_2 > n_5 + n_6$.) It follows that $0 \leq m_7 = n_7 - 3n_5$, so $n_7 \geq 3n_5$, and $n_7 + n_6 - n_2 = n_7 - 2n_5 > 0$: thus the eighth highest value for $\{n_i\}$ must occur at 1123221 rather than 1223210, as the difference is $\alpha_7 + \alpha_6 - \alpha_2$. The corresponding root for $\{m_i\}$ must be either 1223211 or 1122221 – the former gives $n_5 + n_2 = m_6 + m_5 = -\frac{3}{2}n_6 - \frac{3}{2}n_5 + \frac{5}{2}n_2$, so $\frac{3}{2}n_6 + \frac{5}{2}n_5 = \frac{3}{2}n_2 = \frac{3}{2}n_6 + 3n_5$ from above, and this contradicts $n_5 > 0$. Thus we must have 1122221, which gives $n_5 + n_2 = m_5 + m_4 + m_2 = n_6 + 2n_5 + n_4$, so $n_6 + n_5 + n_4 = n_2 = n_6 + 2n_5$ and $n_5 = n_4$.

Now consider the simple roots outside $\langle \alpha_1, \alpha_3, \alpha_4, \alpha_5 \rangle$: the values for $\{n_i\}$ are n_6, n_7 and $n_6 + 2n_5$, while those for $\{m_i\}$ are $n_7 - 3n_5, n_6 + \frac{5}{2}n_5$ and $n_5 + n_6$. The least values in each set of three must agree: the smallest for $\{n_i\}$ clearly cannot be $n_6 + 2n_5$, and it cannot be n_7 as $n_7 - 3n_5$ is smaller than it, so it must be n_6; thus the least value for $\{m_i\}$ must be $n_7 - 3n_5$, and we have $n_7 = n_6 + 3n_5$. The two least values taken outside $\langle \alpha_1, \alpha_3, \alpha_4, \alpha_5 \rangle$ in the two labellings are then n_6 (at α_6 for $\{n_i\}$, at α_7 for $\{m_i\}$), and $n_6 + n_5$ (at $\alpha_6 + \alpha_5$ for $\{n_i\}$, at α_2 for $\{m_i\}$). The next smallest such is $n_6 + 2n_5$ – for $\{n_i\}$ this occurs at $\alpha_6 + \alpha_5 + \alpha_4$ and α_2, and at $\alpha_6 + \alpha_5 + \alpha_4 + \alpha_3$ if $n_3 = 0$, and at $\alpha_6 + \alpha_5 + \alpha_4 + \alpha_3 + \alpha_1$ if $n_3 = n_1 = 0$; for $\{m_i\}$ it occurs at $\alpha_4 + \alpha_2$, and at $\alpha_4 + \alpha_2 + \alpha_3$ if $n_3 = 0$, and at $\alpha_4 + \alpha_2 + \alpha_3 + \alpha_1$ if $n_3 = n_1 = 0$. Thus there is always one more occurrence of $n_6 + 2n_5$ for $\{n_i\}$ than for $\{m_i\}$ – a contradiction. Thus this case cannot occur.

Case B: (i) holds for $\{n_i\}$ and (iii) for $\{m_i\}$, so that $n_6 + n_5 < n_2$ and $m_2 < m_5$. We have $m_2 = n_5$ and $m_5 = n_6 + n_5$: substituting as above gives $m_7 + 2m_6 = n_7 - n_6 - 2n_5 + 2n_2$, $2m_7 + 2m_6 = 2n_7 + 2n_6 + n_5 - n_2$, so $m_7 = n_7 + 3n_6 + 3n_5 - 3n_2$,

$m_6 = -2n_6 - \frac{5}{2}n_5 + \frac{5}{2}n_2$. Exactly as in Case A, the seventh highest value for $\{n_i\}$ occurs at 1123321, and there are the same choices for the corresponding root for $\{m_i\}$: if it is 1223211 we have $n_2 = m_6 + m_5 = -n_6 - \frac{3}{2}n_5 + \frac{5}{2}n_2$, so $\frac{3}{2}n_2 = n_6 + \frac{3}{2}n_5 \leq \frac{3}{2}(n_6 + n_5) < \frac{3}{2}n_2$, a contradiction. Thus it must be 1123221, and $n_2 = m_5 + m_2 = n_6 + 2n_5$ as before. (Note that $n_5 > 0$ as above.) Again, the eighth highest value for $\{n_i\}$ must occur at 1123221 just as in Case A, while the corresponding root for $\{m_i\}$ is 1223211 or 1122221 – the former gives $n_5 + n_2 = m_6 + m_5 = -n_6 - \frac{3}{2}n_5 + \frac{5}{2}n_2$, so $n_6 + \frac{5}{2}n_5 = \frac{3}{2}n_2 = \frac{3}{2}n_6 + 3n_5$ and $\frac{1}{2}n_6 + \frac{1}{2}n_5 = 0$, i.e. $n_6 = n_5 = 0$, a contradiction. Thus we must have 1122221, which gives $n_5 + n_2 = m_5 + m_4 + m_2 = n_6 + 2n_5 + n_4 = n_4 + n_2$, and again $n_5 = n_4$.

Consider simple roots outside $\langle \alpha_1, \alpha_3, \alpha_4, \alpha_5 \rangle$ for $\{n_i\}$, and those outside $\langle \alpha_1, \alpha_2, \alpha_3, \alpha_4 \rangle$ for $\{m_i\}$ (as $m_2 = n_5$, values taken on these subsystems correspond) – the $\{n_i\}$ values are n_7, n_6 and $n_6 + 2n_5$, while the $\{m_i\}$ values are $n_7 - 3n_5$, $\frac{1}{2}n_6 + \frac{5}{2}n_5$ and $n_6 + n_5$. Again, the least values for each must agree: the smallest for $\{n_i\}$ must be n_6 as before, while that for $\{m_i\}$ must be $n_7 - 3n_5$ or $\frac{1}{2}n_6 + \frac{5}{2}n_5$ – thus either $n_7 = n_6 + 3n_5$ or $n_6 = 5n_5$, and we have two subcases.

Subcase B1: If $n_7 = n_6 + 3n_5$, continue to consider values taken outside $\langle \alpha_1, \alpha_3, \alpha_4, \alpha_5 \rangle$ for $\{n_i\}$ and outside $\langle \alpha_1, \alpha_2, \alpha_3, \alpha_4 \rangle$ for $\{m_i\}$. The results are exactly as at the end of Case A (except that $n_6 + n_5$ occurs at α_5 rather than α_2 for $\{m_i\}$), and there is always one more occurrence of $n_6 + 2n_5$ for $\{n_i\}$ than for $\{m_i\}$ – unless this value is also taken at α_6 for $\{m_i\}$, i.e. $n_6 + 2n_5 = \frac{1}{2}n_6 + \frac{5}{2}n_5$, so $n_6 = n_5$, $n_2 = 3n_5$ and $n_7 = 4n_5$.

Now consider the ninth highest value. The maximal roots remaining for $\{n_i\}$ are 1223210, 1123211 and 1122221 – as $n_6 = n_4 = n_5$, the second and third take equal values, and as $n_7 - n_2 = n_5 > 0$, the first takes a smaller value. The corresponding maximal roots for $\{m_i\}$ are 1223211 and 1112221; as $m_6 + m_5 = 5n_5 = n_6 + n_5 + n_2$, the first of these does take the required value. As the roots 1223210 and 1123211 below 1223211 take strictly smaller values for $\{m_i\}$ (since $m_7 = m_2 = n_5 > 0$), 1112221 must also take this value – thus $5n_5 = m_5 + m_4 + m_2 + m_3 = 4n_5 + n_3$, so $n_3 = n_5$.

At this point we can obtain a contradiction by counting the multiplicity of $5n_5$ as a weight outside the A_4 systems having the same labelling.

Subcase B2: If $n_6 = 5n_5 = m_6$, then $n_2 = 7n_5$ and $m_5 = 6n_5$. Since by assumption $m_7 \geq m_6$ we have $n_7 - 3n_5 \geq 5n_5$, so $n_7 \geq 8n_5$. Consider the ninth highest value. The maximal roots remaining for $\{n_i\}$ are as in Subcase B1: as $n_7 - n_2 \geq 8n_5 - 7n_5 = n_5 > 0$, the second takes a higher value than the first, while as $n_6 - n_4 = 5n_5 - n_5 = 4n_5 > 0$, the third takes the highest value. The corresponding maximal roots for $\{m_i\}$ are again as in B1; however as $m_6 + m_5 = 11n_5 > n_5 + n_4 + n_2 = 9n_5$ the first takes too small a value, so the second must be the root concerned and we must have $9n_5 = m_5 + m_4 + m_2 + m_3 = 8n_5 + n_3$, so $n_3 = n_5$.

Now we count the multiplicity of $7n_5$ as a weight outside the A_4 subsystems with common labelling. For this multiplicity to be the same for $\{n_i\}$ and $\{m_i\}$, it is necessary that $m_7 = 7n_5$. But then we find that $10n_5$ has different multiplicities as a weight outside the A_4 systems for $\{n_i\}$ and $\{m_i\}$, a contradiction.

Case C: (ii) holds for $\{n_i\}$ and (iii) for $\{m_i\}$, so that $n_5 \leq n_2 \leq n_5 + n_6$ and $m_2 < m_5$. We have $m_2 = n_5$ and $m_5 = n_2$: substituting as above gives $m_7 + 2m_6 = n_7 + 2n_6 + n_5 - n_2$, $2m_7 + 2m_6 = 2n_7 + 2n_6 + n_5 - n_2$, so $m_7 = n_7$ and $m_6 = n_6 + \frac{1}{2}n_5 - \frac{1}{2}n_2$. Since $m_5 + m_2 = n_5 + n_2$, the root $\alpha' = 1123221 \, (= \alpha - (\alpha_5 + \alpha_2))$ takes equal values under both labellings, as do $1122221 = \alpha' - \alpha_4$, $1112221 = \alpha' - \alpha_4 - \alpha_3$ and $0112221 = \alpha' - \alpha_4 - \alpha_3 - \alpha_1$. Removing these, the four highest roots (which also take equal values) and 1223221 and 1123321 (whose values with $\{m_i\}$ are those with $\{n_i\}$ interchanged, by assumption) leaves a unique maximal root 1223211. Equating values here and subtracting gives $n_6 + n_5 = m_6 + m_5 = n_6 + \frac{1}{2}n_5 + \frac{1}{2}n_2$, so $n_5 = n_2$ and $m_2 = m_5$, a contradiction. Thus this case cannot occur.

Therefore it is not possible for different cases among (i), (ii), (iii) to occur for $\{n_i\}$ and $\{m_i\}$, and as observed above this implies that the labellings are the same.

The cases where $G = F_4$ or G_2 are easy exercises which we leave to the reader. This completes the proof of Theorem 6.3.

Proof of Theorem 6.4

Assume that X_1 and X_2 are closed connected subgroups of the simple exceptional group G such that $X_1 \cong X_2 \cong A_1$ and X_1, X_2 have the same labelled diagram. Let $T_{X_i} = \{T_i(c) : c \in K^*\}$ be a maximal torus of X_i ($i = 1, 2$), where $T_i(c)$ corresponds to the matrix $\mathrm{diag}\,(c, c^{-1})$ in SL_2. Since X_1 and X_2 have the same labelled diagram, replacing X_2 by a suitable Aut G-conjugate, there is a basis $e_\beta \, (\beta \in \Sigma)$, $h_\alpha \, (\alpha \in \Pi)$ of $L(G)$ such that for $i = 1, 2$,

$$T_i(c)e_\beta = c^{l_\beta} e_\beta \quad (\beta \in \Sigma),$$

where the root system Σ is based on a maximal torus T of G containing T_{X_1} and T_{X_2}. In other words, we may take it that $T_1(c) = T_2(c)$ for all $c \in K^*$. Write $T_0 = T_{X_1} = T_{X_2}$. As X_1 and X_2 have the same weights on $L(G)$, it follows that they have the same composition factors on $L(G)$.

The proof of Theorem 6.4 now proceeds by induction on $\dim G$ (with G exceptional) in a series of lemmas. The first lemma shows that the induction starts.

Lemma 6.5 *Theorem 6.4 holds when $G = G_2$.*

Proof. We have $N(A_1, G_2) = 3$, so $p = 0$ or $p > 3$. By Theorem 7, for each i either X_i is a maximal connected subgroup of G, or X_i lies in a subsystem subgroup

$Y = A_1\tilde{A}_1$ or A_2 (where $\tilde{A}_1$ is a short SL_2). If X_i is maximal, then by 2.4, $p \neq 5$ and

$$L(G) \downarrow X_i = 2/10.$$

And if $X_i \leq Y = A_1\tilde{A}_1$ or A_2, then $L(G) \downarrow X_i$ can be determined from

$$L(G) \downarrow Y \;=\; \begin{aligned} &L(A_1)^2/1 \otimes 3, \ \text{if } Y = A_1\tilde{A}_1 \\ &L(A_2)/10/01, \ \text{if } Y = A_2. \end{aligned}$$

If X_1 and X_2 are either both maximal, or both lie in conjugates of the same subsystem group Y, then X_1 and X_2 are G-conjugate. Otherwise, we have $X_1 < A_2$ and $X_2 < A_1\tilde{A}_1$ (or vice versa); but then $X_1 = C_{A_2}(t)$ for some involution $t \in N_G(A_2)$, so $X_1 \leq C_G(t)^0 = A_1\tilde{A}_1$, whence X_1 and X_2 are again conjugate. $\square$

Lemma 6.6 *Suppose that there is a proper closed connected semisimple subgroup D of G containing X_1 and X_2. Then X_1 and X_2 are Aut G-conjugate.*

Proof. Write $D = R_1 \ldots R_k$, a commuting product of simple factors R_i, and take D to be minimal subject to containing X_1 and X_2. For a subgroup A of D, write $\bar{A} = AZ(D)/Z(D)$, so that $\bar{D} = \bar{R}_1 \times \ldots \times \bar{R}_k$. Let $\pi_i : \bar{D} \to \bar{R}_i$ be the ith projection map. Choose an isomorphism $x_1 \to x_2$ from X_1 to X_2 which fixes $T_1(c)$ for all c. Also, let $\tilde{D}$ be the simply connected cover of D, and let $\tilde{X}_1, \tilde{X}_2$ be the connected preimages of X_1, X_2 in $\tilde{D}$. For $i = 1, 2$ choose a rank 1 torus $\tilde{T}_i = \{\tilde{T}_i(c) : c \in K^*\}$ of $\tilde{X}_i$ such that $\tilde{T}_i(c)$ maps to $T_i(c)$ for all c.

Suppose first that no factor R_i is E_6 or D_m $(m \geq 4)$. Then for all i, either $R_i \cong A_m$ or Aut $R_i \cong R_i/Z(R_i)$. We claim that for each i, there exists $r_i \in R_i$ such that

$$\bar{x}_1\pi_i = (\bar{x}_2\pi_i)^{\bar{r}_i} \quad \text{for all } x_1 \in X_1.$$

It then follows that X_1 and X_2 are conjugate by the element $r_1 \ldots r_k$ of D, giving the conclusion. If R_i is classical of type A_m, B_m or C_m, with natural module V_i, then $\tilde{T}_1(c)^{V_i} = \pm\tilde{T}_2(c)^{V_i}$ for all c. This implies that $\tilde{T}_1(c^2)^{V_i} = \tilde{T}_2(c^2)^{V_i}$ for all c, so in fact $\tilde{T}_1(d)^{V_i} = \tilde{T}_2(d)^{V_i}$ for all d. It follows that $\tilde{X}_1$ and $\tilde{X}_2$ have the same weights on V_i, hence have the same composition factors on V_i. By our assumption on p, $V_i \downarrow \tilde{X}_1$ and $V_i \downarrow \tilde{X}_2$ are completely reducible (see [AJL, 3.9]), so the representations of $\tilde{X}_1$ and $\tilde{X}_2$ on V_i are equivalent. The existence of the element r_i follows easily (the possibility that $R_i = D_m$ was excluded here since in this case a graph automorphism of R_i may be required to conjugate $\tilde{X}_1^{V_i}$ to $\tilde{X}_2^{V_i}$). Now suppose that R_i is exceptional of type G_2, F_4 or E_7. The fact that $T_1(c) = T_2(c)$ means that $\bar{X}_1\pi_i$ and $\bar{X}_2\pi_i$ have the same labelled diagram as subgroups of $\bar{R}_i$, and hence the existence of r_i follows by induction.

Now suppose that some R_i, say R_1, is E_6 or D_m $(m \geq 4)$. If $R_1 = E_6$ then $D = E_6 A_r$ with $0 \leq r \leq 2$ (interpreting A_0 as 1), and by Theorem 5, R_1 is a subsystem subgroup of G. Hence $N_G(D)$ induces a graph automorphism of R_1, and so by induction there is an element $\alpha_1 \in N_G(D)$ such that $\bar{x}_1\pi_1 = (\bar{x}_2\pi_1)^{\alpha_1}$ for all $x_1 \in X_1$. The result now follows as before. If $R_1 = D_m$ with $m \geq 5$, the same argument applies (note that Aut G contains a 2-element inducing a graph automorphism on D_m).

It remains to consider the case where $R_1 = D_4$. The fact that $T_1(c) = T_2(c)$ for all c forces $\tilde{X}_1$ and $\tilde{X}_2$ to have equivalent representations on each of the 8-dimensional modules $V_{R_1}(\lambda_i)\,(i = 1, 3, 4)$. By considering the effect of triality, we see that for some i, $V_{R_1}(\lambda_i)$ restricts reducibly to both $\tilde{X}_1$ and $\tilde{X}_2$. This puts $\bar{X}_1\pi_1$ and $\bar{X}_2\pi_1$ in conjugates of B_rB_{3-r} for some r, and hence we see that $\bar{X}_1\pi_1$ and $\bar{X}_2\pi_1$ are $\bar{R}_1$-conjugate. It follows as before that X_1 and X_2 are D-conjugate. $\square$

Lemma 6.7 *Suppose that either $p = 0$, or $p > 0$ and $L(G) \downarrow X_1$ has no composition factor of the form $(p-2) \otimes c^{(p)}$ with $p \nmid c$. Then X_1 and X_2 are Aut G-conjugate.*

Proof. Write $[xy]$ for the Lie product on $L(G)$. Let $L(T_0) = \langle h \rangle$, and write

$$L(X_1) = \langle e, h, f \rangle, \quad L(X_2) = \langle e', h', f' \rangle$$

where these are the usual bases for sl_2 (so $[he] = 2e, [hf] = -2f, [ef] = h$ and similarly for e', h', f'). Let $V = L(G)$, and for any integer i define V_i to be the i-weight space for T_0 on V. Then the function

$$\phi : v \to [ve] \quad (v \in V_0)$$

is a linear map from V_0 to V_2.

We claim that $V_0\phi = V_2$. This is clear from the representation theory of SL_2 if $p = 0$, so suppose $p > 0$. To prove the claim, let

$$V = W_0 \supset W_1 \supset \ldots \supset W_k = 0$$

be an X_1-composition series for V. The irreducible X_1-module V/W_1 is isomorphic to $c_0 \otimes c_1^{(p)} \otimes \ldots \otimes c_r^{(p^r)}$ with $0 \leq c_i \leq p-1$ for all i; say $V/W_1 = A_0 \otimes A_1 \otimes \ldots \otimes A_r$ with $A_i \cong c_i^{(p^i)}$. If $x \in V/W_1$ is a non-zero vector of T_0-weight 2, then either

(i) $x = a_0 \otimes y$, where $a_0 \in A_0$ has T_0-weight 2 and T_0 fixes y, or

(ii) $x = a_0 \otimes y$, where $a_0 \in A_0$ has T_0-weight $-p+2$ and y has T_0-weight p.

In case (ii), $V/W_1 \cong (p-2) \otimes c^{(p)}$ with $p \nmid c$, contrary to hypothesis. Hence (i) holds. Then $a_0 = ae$ for some $a \in A_0$ fixed by T_0, so $x = [(a \otimes y)e]$, where $a \otimes y$ is fixed by T_0. Therefore

$$V_2 + W_1 = [V_0e] + W_1.$$

Repeating the argument successively for $W_1, \ldots, W_k$, gives $V_2 = [V_0e] = V_0\phi$, as claimed.

Now define

$$P = \langle T, U_{\pm\beta} : l_\beta \geq 0 \rangle$$

(recall that T was chosen as a maximal torus of G containing T_0). By [Se2, 2.4], P is a parabolic subgroup of G, with unipotent radical $Q = \langle U_\beta : l_\beta > 0 \rangle$ and Levi subgroup $R = C_G(T_0)$; moreover, e and e' lie in $L(Q)$ (see [Se2, 2.4(iii)]). Since we know that $V_2 = [V_0 e]$, we have

$$\dim C_{V_0}(e) = \dim V_0 - \dim V_2.$$

As $V_0 = L(R)$, this gives $\dim L(R) - \dim C_{L(R)}(e) = \dim V_2$, and hence

$$\dim R - \dim C_R(e) \geq \dim V_2.$$

In other words, the dimension of the R-orbit containing e is at least $\dim V_2$. This implies that e lies in the unique dense orbit of R on V_2. Similarly e' lies in the dense orbit, and so e and e' are R-conjugate. Hence we may assume that $e = e'$, and so

$$L(X_1) = \langle e, h, f \rangle, \quad L(X_2) = \langle e, h, f' \rangle$$

with $[ef] = [ef'] = h$ and $[he] = 2e$. Suppose $f - f' \neq 0$, and write $l = f - f'$. Then $l \in C_V(e) \cap V_2$ and $p > 0$. Choose an X_1-composition factor H/K of V with $l \in H - K$, and write

$$H/K = B_0 \otimes B_1 \otimes \ldots \otimes B_k$$

with $B_i \cong d_i^{(p^i)}$ for $0 \leq i \leq k$ (where $0 \leq d_i \leq p-1$). Then $C_{H/K}(e) = \langle b \rangle \otimes B_1 \otimes \ldots \otimes B_k$, where b is a high weight vector in B_0. Since $l + K \in C_{H/K}(e)$ has T_0-weight -2, we must have $d_0 = p - 2$ and $d_1 \geq 1$. But then $H/K \cong (p-2) \otimes c^{(p)}$ with $p \nmid c$, contrary to hypothesis. Therefore $f - f' = 0$.

We have now established that $L(X_1) = L(X_2)$. Write $C = C_G(L(X_1))^0$. Then X_1 and X_2 both lie in $N_G(L(X_1))^0 = X_1 C$. If C is reductive then $X_1 C$ is reductive, and the result follows by 6.6. Otherwise, $R_u(C) \neq 1$ and so $X_1 C$ lies in a parabolic subgroup $P_0 = Q_0 R_0$ with unipotent radical Q_0 and Levi subgroup R_0. By Theorem 1, there exist $u, v \in Q_0$ such that X_1^u and X_2^v lie in R_0. Then $T_1(c)^u = T_2(c)^v$ for all c, and it follows that X_1^u and X_2^v give the same labelled diagram of R_0. Now 6.6 gives the required conclusion. □

In view of 6.7, we assume from now on that $p > 0$ and $L(G) \downarrow X_1$ has a composition factor $(p-2) \otimes c^{(p)}$ with $p \nmid c$.

Lemma 6.8 *Theorem 6.4 holds when $G = F_4$.*

Proof. Here $p \geq 5$, by assumption. By Theorem 7, either X_1 lies in a proper subsystem subgroup Y of G, or X_1 lies essentially in a maximal connected subgroup A_1 $(p \geq 13)$, $A_1 G_2$ $(p \geq 7)$ or G_2 $(p = 7)$.

Suppose first that $X_1 \leq Y$, a proper subsystem subgroup, and choose Y to be minimal such. The possibilities for Y are

$$A_1, \tilde{A}_1, A_1^2, A_1 \tilde{A}_1, A_2, \tilde{A}_2, B_2, A_1^3, A_1^2 \tilde{A}_1, A_3, B_2 A_1, C_3,$$

$$B_3, \tilde{A}_2 A_1, A_2 \tilde{A}_1, A_1^4, A_2 \tilde{A}_2, A_3 \tilde{A}_1, C_2 A_1 A_1, C_3 A_1, D_4, B_4,$$

where $\tilde{A}_1, \tilde{A}_2$ denote short root subsystems (see [Ca, Table 8]). All of these embed in $B_4, A_1 C_3$ or $A_2 \tilde{A}_2$, so the composition factors of $L(G) \downarrow Y$ can be read off using 2.1; and for each factor $V_Y(\lambda)$ occurring, the composition factors of $V_Y(\lambda) \downarrow X_1$ are given by 2.13. From this we see that $L(G) \downarrow X_1$ can have a composition factor of the form $(p-2) \otimes c^{(p)}$ with $p \nmid c$ only in the following cases:

$$p = 5 : Y = B_2 A_1, A_3 \tilde{A}_1, C_2 A_1 A_1, C_3 A_1, D_4, B_4$$
$$p = 7 : Y = C_2 A_1 A_1, C_3 A_1$$
$$p = 11 : Y = C_3 A_1.$$

If Y has more than one factor, then except for one case, one of the projections of X_1 to the factors involves a Frobenius p-power twist; in the exceptional case, $p = 5$ or 7, $Y = C_3 A_1$ and the projection $X_1 \to C_3$ corresponds to the representation $1^{(p)} \otimes 2$ or $1 \otimes 2^{(p)}$.

Assume now that $p \geq 11$. Then by the above, either X_1 and X_2 both lie in conjugates of a subsystem subgroup $C_3 A_1$, or both lie in conjugates of a maximal subgroup $A_1 G_2$. In either case we see that X_1 and X_2 must be G-conjugate.

Suppose now that $p = 5$. From $L(G) \downarrow Y$ we see that the highest weight of T_0 on $L(G)$ is as in the following table. We also specify the embedding of X_1 in Y by giving the restrictions to X_1 of the natural modules for the factors of Y. In the table, q, r denote powers of p with $q > 1, r \geq 1$.

Y	embedding of X_1	highest T_0-weight
$B_2 A_1$	$4, 1^{(5)}$	9
$A_3 \tilde{A}_1$	$3, 1^{(5)}$	14
$C_2 A_1 A_1$	$3, 1^{(5)}, 1^{(r)}$	$10(r=1); 14(r=5); 2r(r>5)$
$C_3 A_1$	$1 \otimes 2^{(q)}, 1^{(5)}$	$6 + 4q$
	$1^{(5)} \otimes 2, 1$	16
D_4	$3 \otimes 1^{(5)}$	14
B_4	$2 \otimes 2^{(5)}$	22

Note that since $p = 5$, X_1 does not lie in a maximal connected non-subsystem subgroup of G. Also, when X_1 lies in $C_2 A_1 A_1$ with embedding $3, 1^{(5)}, 1^{(5)}$, then $X_1 < C_2 \tilde{A}_1 < A_3 \tilde{A}_1$.

From the above we conclude that when $p = 5$, either X_1 and X_2 lie in conjugates of the same subgroup Y and are G-conjugate, or $X_1 < A_3 \tilde{A}_1$ and $X_2 < D_4$. In the latter case, a conjugate of X_2 lies in a subgroup $B_2 B_1$ of D_4 (since $N_G(D_4)$ induces a group S_3 of graph automorphisms on D_4), and hence X_2 lies in a subsystem subgroup $A_3 \tilde{A}_1$ (lying in a subsystem B_4 containing D_4). Hence X_1 and X_2 are G-conjugate.

When $p = 7$, either $X_1 < Y = C_2 A_1 A_1$ or $C_3 A_1$, or X_1 lies in a maximal subgroup G_2 (if $X_1 < A_1 G_2$ there is no factor $(p-2) \otimes c^{(p)}$). Moreover, if $X_1 < C_3 A_1$ then the embedding is $1^{(7)} \otimes 2, 1$, since $L(G) \downarrow X_1$ has a factor $5 \otimes c^{(7)}$. Hence, comparing highest T_0-weights, we see that X_1 and X_2 both lie in a conjugate of $C_2A_1A_1$, a conjugate of C_3A_1, or a conjugate of G_2, so are G-conjugate. $\square$

Lemma 6.9 *Theorem 6.4 holds when $G = E_6$.*

Proof. Here $p \geq 7$. By Theorem 7, X_1 lies either in a proper subsystem subgroup Y, or in a maximal subgroup $Z = F_4, C_4, G_2(p \geq 11), A_2$ or A_2G_2 of G.

Suppose $X_1 \leq Z$, a maximal non-subsystem subgroup. Then $L(G) \downarrow Z$ is given by 2.4. If $Z = C_4$ then $V_Z(\lambda_1) \downarrow X_1$ is either 7 or $1 \otimes 1^{(q)} \otimes 1^{(q')}$. In the first case, $p \geq 11$ and T_0 has highest weight 16 on $L(G)$ by [Se2, p.65], so there is no factor $(p-2) \otimes c^{(p)}$; and in the second case we restrict from $L(G) \downarrow Z$ to X_1 to see again that no factor $(p-2) \otimes c^{(p)}$ occurs. When $Z = G_2(p \geq 11)$, $L(G) \downarrow Z = L(Z)/11$ and the highest T_0-weight on $L(G)$ is 16 again. The cases $Z = A_2$ or A_2G_2 are excluded similarly. Finally, if $Z = F_4$ then X_1 lies in a maximal subgroup $A_1(p \geq 13), G_2(p = 7)$ or A_1G_2 of Z. In the first case the highest T_0-weight is 22 by [Se2, p.65]; and in the second, $X_1 < A_1G_2 < A_2G_2$, a case considered before. This leaves the case where $X_1 < G_2 < Z = F_4$ $(p = 7)$.

Now suppose $X_1 < Y$, a subsystem group. Then Y lies in a subsystem group A_1A_5, A_2^3 or D_5. The restrictions of $L(G)$ to the first two of these subgroups are given by 2.1, and $L(G) \downarrow D_5 = L(D_5)/\lambda_4/\lambda_5/0$. Hence we see that $(p-2) \otimes c^{(p)}$ can only occur in $L(G) \downarrow X_1$ if $Y = A_1A_5$, $p = 7$ or 11 and the embedding of X_1 in Y is given by the representations $1^{(p)}, 5$ or $1, 1^{(p)} \otimes 2$. If also $X_2 < A_1A_5$, then the embedding of X_2 must be given by the same representations as for X_1 (compare highest T_0-weights), so X_1 and X_2 are G-conjugate.

We conclude that either X_1 and X_2 are G-conjugate, or $p = 7$, $X_1 < A_1A_5$ and $X_2 < G_2$ (with G_2 maximal in a subgroup F_4 of G). Comparison of the highest T_0-weights shows that the embedding $X_1 < A_1A_5$ is given by the representations $1^{(7)}, 5$. Then X_1 lies in a subgroup A_1C_3 of A_1A_5, and this A_1C_3 lies in a subgroup F_4 of G. All subgroups of type F_4 are G-conjugate, so there exist $g \in G$ and a subgroup F_4 such that X_1^g and X_2 both lie in F_4. Moreover, one checks using 2.1 and 2.4 that X_1^g and X_2 have the same composition factors on $L(F_4)$, so by 6.3 they have the same labelled diagrams as subgroups of F_4. Therefore they are F_4-conjugate by 6.8. $\square$

In view of 6.8 and 6.9, we suppose from now on that G is E_7 or E_8 and $p \geq 11$ (as $N(A_1, E_7) = 7$). The next lemma locates those subgroups X_1 for which $L(G) \downarrow X_1$ has a composition factor $(p-2) \otimes c^{(p)}$.

Let Y be a subsystem subgroup of $G = E_7$ or E_8, minimal subject to containing X_1 (possibly $Y = G$, of course).

Lemma 6.10 *The possibilities for Y are as follows:*

$$G = E_7 : \quad Y = A_1A_5 \text{ or } A_1D_6;$$
$$G = E_8 : \quad Y = A_1A_5, A_1^2A_5, A_1A_2A_5, A_1^2D_5, A_3D_5, A_1D_6,$$
$$A_1^2D_6, D_4D_4, D_7, D_8 \text{ or } A_1E_7.$$

Proof. Suppose first that Y is a product of classical groups, so $Y = Y_1 \ldots Y_k$ with each Y_i a simple group of type A or D. For each i, let $X_1^{(i)}$ be the projection of X_1 in Y_i, and let $V_i = V_{Y_i}(\lambda_1)$. Write $\tilde{X}_1^{(i)}$ for the connected preimage of $X_1^{(i)}$ in the simply connected cover of Y_i. By the minimality of Y, each $X_1^{(i)}$ is essentially embedded in Y_i. Thus if $Y_i = A_m$ then $\tilde{X}_1^{(i)}$ is irreducible on V_i and the possibilities for $V_i \downarrow \tilde{X}_1^{(i)}$ are listed in 2.9 (since $\dim V_i \leq 9$). And if $Y_i = D_m$ $(m \geq 4)$ then $\tilde{X}_1^{(i)}$ is either irreducible on V_i, or embeds in a subgroup $B_r B_{m-r-1}$ of Y_i, irreducible in each factor.

Now if $Y_i = A_m$ then the nontrivial composition factors of $(L(G)/L(Y_i)) \downarrow Y_i$ have high weights $\lambda_1, \lambda_2, \lambda_3$ (and λ_4 if $Y_i = A_7'$), together with duals of these; and if $Y_i = D_m$ the composition factors have high weights λ_1, λ_{m-1} and λ_m. Thus the high weights of composition factors of $L(G) \downarrow X_1^{(i)}$ can be read off using 2.13. In the table on the following page, we list various relevant high weights occurring for each possible Y_i apart from D_7 and D_8. In the table, q, q', r are powers of p with $q, q' > 1$ and $r \geq 1$. If a composition factor $a \otimes b^{(q)}$ is listed, it is understood that $b \otimes a^{(q)}$ is also included. When we write '...' after a list of weights, this indicates that the only further high weights occurring are smaller than those listed, and cannot possibly contribute to a composition factor $(p-2) \otimes c^{(p)}$ for X_1.

The composition factors of $L(G) \downarrow X_1$ are factors of $L(Y) \downarrow X_1$, together with tensor products of twists of composition factors of $L(G) \downarrow X_1^{(i)}$ $(i = 1, \ldots, k)$. We know that $(p-2) \otimes c^{(p)}$ $(p \geq 11, p \nmid c)$ occurs as a composition factor of $L(G) \downarrow X_1$. The possibilities for the subsystem group Y are listed in [Ca, Tables 10,11]. A check using the above table of high weights occurring in $(L(G)/L(Y_i)) \downarrow X_1^{(i)}$ quickly excludes all the subsystem groups in [Ca] except those listed in the conclusion of this lemma, together with $A_2 A_5$ and $A_1 A_7$. However, from 2.1 and 2.3 we see that the only composition factors of $L(G) \downarrow A_2 A_5$ which are tensor products of nontrivial modules for both factors are of the form $\lambda_1 \otimes \lambda_1$, $\lambda_1 \otimes \lambda_2$ and duals of these. Hence using 2.13, we see that $(p-2) \otimes c^{(p)}$ cannot occur in $L(G) \downarrow X_1$ when $Y = A_2 A_5$. Similarly $Y \neq A_1 A_7$.

This completes the proof when Y is a product of classical groups. Suppose now that X_1 lies in no such subsystem group. Then by Theorem 7, there is a subgroup $Y_0 = F_4, E_6, E_7$ or E_8 of G, a maximal connected subgroup Z of Y_0 not containing a maximal torus, and a semisimple subgroup Y_1 of $C_G(Y_0)$, such that X_1 is subdirectly maximal in $Y_1 Z$. Let $X_1^{(0)}$ be the projection of X_1 in Z. The possibilities for $L(G) \downarrow Y_1 Z$ can be read off from 2.1, 2.4 and 2.5. We see from this that the high

Y_i	$V_i \downarrow \tilde{X}_1^{(i)}$	comp. factors in $(L(E_8)/L(Y_i)) \downarrow X_1^{(i)}$
A_1	1	1
A_2	2	2
A_3	3	$4, 3, 2$
	$1 \otimes 1^{(q)}$	$1 \otimes 1^{(q)}, 2, 2^{(q)}$
A_4	4	$6, 4, 2, 0$
A_5	5	$9, 8, 5, 4, \dots$
	$2 \otimes 1^{(q)}$	$4 \otimes 1^{(q)}, 2 \otimes 2^{(q)}, 4, 3^{(q)}, \dots$
A_6	6	$12, 10, 8, 6, \dots$
A_7	7	$16, 15, 12, 11, 8, 7, \dots$
	$3 \otimes 1^{(q)}$	$6 \otimes 2^{(q)}, 7 \otimes 1^{(q)}, 8, \dots$
	$1 \otimes 1^{(q)} \otimes 1^{(q')}$	$3 \otimes 1^{(q)} \otimes 1^{(q')}, 4^{(q)}, 4^{(q')}, \dots$
A_8	8	$18, 14, 10, \dots$
	$2 \otimes 2^{(q)}$	$4 \otimes 4^{(q)}, 6^{(q)}, \dots$
D_4	$3 \otimes 1^{(q)}$	$3 \otimes 1^{(q)}, 4, 2^{(q)}, \dots$
	$6/0$	$6, \dots$
D_5	$8/0$	$10, 8, 4, \dots$
	$6/2^{(r)}$	$6 \otimes 1^{(r)}, \dots$
	$4/4^{(r)}$	$3 \otimes 3^{(r)}, 4^{(r)}, 4, \dots$
D_6	$5 \otimes 1^{(q)}$	$9, 8 \otimes 1^{(q)}, 5 \otimes 2^{(q)}, 5 \otimes 1^{(q)}, 3^{(q)}, \dots$
	$10/0$	$15, 10, 9, 5, \dots$
	$8/2^{(r)}$	$10 \otimes 1^{(r)}, 8, \dots$
	$6/4^{(r)}$	$6 \otimes 3^{(r)}, 6, \dots$

weights of composition factors of $L(G) \downarrow X_1^{(0)}$ are as in the following table (again, q, r denote powers of p with $q > 1, r \geq 1$; and a superscript (r) by a simple factor of Z indicates a Frobenius r-power twist in the projection of $X_1^{(0)}$ to that factor; also, if a high weight $a \otimes b^{(r)}$ is written, it is understood that $b \otimes a^{(r)}$ is also possible).

Y_0	Z	high wts. in $(L(Y_0)/L(Z)) \downarrow X_1^{(0)}$	high wts. in $(L(E_8)/L(Y_0)) \downarrow X_1^{(0)}$
E_8	$A_1, p \geq 23$	$38, \ldots$	
	$A_1, p \geq 29$	$46, \ldots$	
	$A_1, p \geq 31$	$58, \ldots$	
	B_2	$18, \ldots$	
	$A_1^{(r)} A_2$	$6^{(r)} \otimes 4, 2^{(r)} \otimes 8, \ldots$	
	$G_2^{(r)} F_4, p \geq 13$	$6^{(r)} \otimes 16, 6^{(r)} \otimes 8$	
E_7	$A_1, p \geq 17$	$26, 22, \ldots$	$21, 15, \ldots$
	$A_1, p \geq 19$	$34, 26, \ldots$	$27, 17, \ldots$
	A_2	$16, \ldots$	$12, \ldots$
	$A_1^{(r)} A_1$	$2^{(r)} \otimes 8, 4^{(r)} \otimes 6, \ldots$	$6^{(r)} \otimes 3, \ldots$
	$A_1^{(r)} G_2$	$4^{(r)} \otimes 6, 2^{(r)} \otimes 12, 2^{(r)} \otimes 8, \ldots$	$1^{(r)} \otimes 10, 3^{(r)} \otimes 6, \ldots$
	$A_1^{(r)} F_4, p \geq 13$	$2^{(r)} \otimes 16, 2^{(r)} \otimes 8, \ldots$	$1^{(r)} \otimes 16, 1^{(r)} \otimes 8, \ldots$
	$G_2^{(r)} C_3$	$6^{(r)} \otimes 8, 6^{(r)} \otimes 2 \otimes 2^{(q)}, 6^{(r)} \otimes 4^{(q)}, \ldots$	$6^{(r)} \otimes 5, 6^{(r)} \otimes 2 \otimes 1^{(q)}, 9, \ldots$
E_6	A_2	$10, \ldots$	$8, \ldots$
	G_2	$16, \ldots$	$12, 8, \ldots$
	$F_4, p \geq 13$	$16, \ldots$	$16, 8, \ldots$
	C_4	$16, 4^{(q)}, 4^{(q')}, \ldots$	$12, 8, 2^{(q)} \otimes 2^{(q')}, \ldots$
	$A_2^{(r)} G_2$	$4^{(r)} \otimes 6, 4^{(r)}, \ldots$	$2^{(r)} \otimes 6, 4^{(r)}, \ldots$
F_4	$A_1, p \geq 13$	$22, \ldots$	$16, 8, \ldots$
	$A_1^{(r)} G_2$	$4^{(r)} \otimes 6, \ldots$	$2^{(r)} \otimes 6, \ldots$

The composition factor $(p-2) \otimes c^{(p)}$ in $L(G) \downarrow X_1$ either occurs in $L(Y_0) \downarrow X_1^{(0)}$, or is a tensor product of factors in $L(G) \downarrow Y_1$ and $(L(E_8)/L(Y_0)) \downarrow X_1^{(0)}$. Hence we see that either $Y_1 Y_0 = A_1 E_7$, or $Y_1 Y_0 = A_1 A_1 F_4 < G_2 F_4$. In the latter case $Y_1 Y_0 < A_1 E_7$ anyway.

This completes the proof of the lemma. $\square$

Lemma 6.11 *If Y is as in 6.10, then $(L(G)/L(Y)) \downarrow Y$ is as follows:*

Y	$(L(G)/L(Y)) \downarrow Y$
$A_1 A_5$	$1 \otimes \lambda_3 / 1 \otimes \lambda_1 / 1 \otimes \lambda_5 / 0 \otimes \lambda_2 / 0 \otimes \lambda_4 / 0$, *if* $G = E_7$
	$1 \otimes \lambda_3 / (1 \otimes \lambda_1)^3 / (1 \otimes \lambda_5)^3 / (0 \otimes \lambda_2)^3 / (0 \otimes \lambda_4)^3 / 0^8$, *if* $G = E_8$
$A_1 D_6$	$1 \otimes \lambda_5$, *if* $G = E_7$
	$1 \otimes \lambda_6 / (1 \otimes \lambda_1)^2 / (0 \otimes \lambda_5)^2 / 0^3$, *if* $G = E_8$
$A_1 A_1 A_5$	$1 \otimes 1 \otimes \lambda_1 / 1 \otimes 1 \otimes \lambda_5 / 0 \otimes 1 \otimes \lambda_1 / 0 \otimes 1 \otimes \lambda_5 / 1 \otimes 0 \otimes \lambda_2 /$
	$1 \otimes 0 \otimes \lambda_4 / 0 \otimes 1 \otimes \lambda_3 / 0 \otimes 0 \otimes \lambda_2 / 0 \otimes 0 \otimes \lambda_4 / (1 \otimes 0 \otimes 0)^2 / 0$
$A_1 A_2 A_5$	$1 \otimes 10 \otimes \lambda_1 / 1 \otimes 01 \otimes \lambda_5 / 1 \otimes 00 \otimes \lambda_3 / 0 \otimes 10 \otimes \lambda_2 / 0 \otimes 01 \otimes \lambda_4$
$A_1 A_1 D_5$	$1 \otimes 1 \otimes \lambda_1 / 1 \otimes 0 \otimes \lambda_4 / 1 \otimes 0 \otimes \lambda_5 / 0 \otimes 1 \otimes \lambda_4 /$
	$0 \otimes 1 \otimes \lambda_5 / (0 \otimes 0 \otimes \lambda_1)^2 / (1 \otimes 1 \otimes 0)^2 / 0$
$A_3 D_5$	$010 \otimes \lambda_1 / 100 \otimes \lambda_4 / 001 \otimes \lambda_5$
$A_1 A_1 D_6$	$1 \otimes 1 \otimes \lambda_1 / 1 \otimes 0 \otimes \lambda_5 / 0 \otimes 1 \otimes \lambda_6$
$D_4 D_4$	$\lambda_1 \otimes \lambda_1 / \lambda_3 \otimes \lambda_3 / \lambda_4 \otimes \lambda_4 /$
D_7	$\lambda_1^2 / \lambda_6 / \lambda_7 / 0$
D_8	λ_7
$A_1 E_7$	$1 \otimes \lambda_7$

Proof. This follows easily from 2.1 and 2.3. □

Lemma 6.12 *The embedding of X_1 in Y, and the highest weight of T_0 on $L(G)$, are as in the table on the following page. For $Y \neq A_1 E_7$, the table describes the embedding of X_1 by giving the restriction to X_1 of the natural modules for the factors of Y (and always q, r, s, t are powers of p with $q > 1$ and $r, s, t \geq 1$); and for $Y = A_1 E_7$, X_1 is essential in $A_1 Z$, where Z is a maximal connected subgroup of E_7, as indicated in the table.*

Proof. The composition factors of $L(G) \downarrow Y$ are given in 6.11, from which those of $L(G) \downarrow X_1$ for the various possible embeddings $X_1 < Y$ can be deduced using 2.13. It is routine to check that the only embeddings $X_1 < Y$ for which $L(G) \downarrow X_1$ has a composition factor $(p-2) \otimes c^{(p)}$ are those listed, with highest T_0-weights as given; note that we have omitted the cases where $Y = A_1 E_7$ and $X_1 < A_1^{(p)} A_1 A_1$, $A_1^{(p)} A_1 G_2$ or $A_1^{(p)} G_2 C_3$ (with representation 5 in C_3), as in these cases X_1 lies in $A_1 A_2 A_5$, $A_1 A_1 D_6$ or $A_1 A_1 D_6$, respectively, for the following reasons. First consider $X_1 < A_1^{(p)} A_1 A_1 < A_1 E_7$. Let $X_1^{(0)}$ be the projection of X_1 in the E_7 factor. By 2.4, we have

$$L(E_7) \downarrow X_1^{(0)} = 2/2/2 \otimes 8/4 \otimes 6/6 \otimes 4/2 \otimes 4/4 \otimes 2 = 10^3/8^3/6^5/4^4/2^6.$$

Y	embedding of X_1	p	highest wt. of T_0
A_1A_5	$1^{(p)}, 5$	11	$2p$
A_1D_6	$1^{(p)}, 5 \otimes 1^{(q)}$	11	$2q + p + 5$
	$1, 5 \otimes 1^{(p)}$	11	$3p + 1$
	$1^{(p)}, 10/0$	11, 17	$p + 15\,(p = 11); 2p\,(p = 17)$
	$1, 8/2^{(p)}$	13	$2p + 8$
	$1^{(p)}, 8/2$	13	$2p$
	$1^{(p)}, 6/4$	11	$2p$
$A_1A_1A_5$	$1^{(r)}, 1^{(p)}, 5$	11	$2p\,(r = 1); 2p + 5\,(r = p); 2r\,(r > p)$
$A_1A_2A_5$	$1^{(p)}, 2^{(r)}, 5$	11	$2p\,(r = 1); 4r\,(r \geq p)$
$A_1A_1D_5$	$1, 1^{(p)}, 8/0$	11	$2p$
A_3D_5	$3, 6/2^{(p)}$	11	$2p + 6$
	$1 \otimes 1^{(p)}, 8/0$	13	$2p + 8$
$A_1A_1D_6$	$1^{(r)}, 1^{(p)}, 10/0$	11, 13, 17	$2p, 2r, p + 15$ or $p + r + 10$
	$1^{(r)}, 1^{(s)}, 8/2^{(t)}$	11, 13	$2r, 2s, 2t + 8, r + s + 8, r + s + 2t,$
			$r + t + 10$ or $s + t + 10$
	$1^{(r)}, 1^{(p)}, 6/4$	11	$2p\,(r = 1); 2p + 6\,(r = p); 2r\,(r > p)$
	$1^{(r)}, 1^{(s)}, 5 \otimes 1^{(q)}$	11	$2r, 2s, 2q + r + 5, 2q + s + 5, 3q + r$ or $3q + s$
D_4D_4	$3 \otimes 1^{(p)}, 6/0$	11	$2p + 6$
D_7	$10/2^{(p)}$	11, 17	$2p + 10$
D_8	$7 \otimes 1^{(p)}$	11, 13, 17	$3p + 7$
	$4 \otimes 2^{(p)}/0$	11	$5p + 3$
	$12/2^{(p)}$	13, 17, 23	$2p + 12$
	$10/4^{(p)}$	11, 17	$4p + 10$
	$2 \otimes 2^{(p)}/6$	11	$4p + 2$
A_1E_7	$X_1 < A_1^{(p)}A_1, p \geq 17$	17, 23	$p + 21\,(p = 17); 2p\,(p = 23)$
	$X_1 < A_1^{(p)}A_1, p \geq 19$	19, 29	$p + 27\,(p = 19); 2p\,(p = 29)$
	$X_1 < A_1A_1^{(p)}G_2$	13	$4p + 6$
	$X_1 < A_1^{(p)}A_1F_4$	17, 19	$2p$
	$X_1 < A_1A_1^{(p)}F_4$	17, 19	$3p + 1$
	$X_1 < A_1^{(p)}G_2^{(q)}C_3$	11	$10q$
	$X_1 < A_1G_2C_3$	11	$3p + 1$
	(rep. $2 \otimes 1^{(p)}$ in C_3)		

Now let $X_0 < A_2A_5 < E_7$, with $X_0 \cong A_1$ diagonally embedded in A_2A_5 via the representations 2,5. Using 2.1, we see that $L(E_7) \downarrow X_0$ has the same composition factors as $L(E_7) \downarrow X_1^{(0)}$. Then by 6.3, X_0 and $X_1^{(0)}$ have the same labelled E_7-diagram. Hence by 6.7, $X_1^{(0)}$ is E_7-conjugate to X_0, and so $X_1^{(0)} < A_2A_5$. Therefore $X_1 < A_1A_2A_5$, as claimed. The other two cases are entirely similar. $\square$

Lemma 6.13 *Suppose that X_1 and X_2 lie in conjugates Y^g, Y^h of the same subgroup Y, with both embeddings $X_1 < Y^g$ and $X_2 < Y^h$ as given in a single row of the table in 6.12. Then X_1 and X_2 are G-conjugate.*

Proof. From the table in 6.12, it is clear that $X_1^{g^{-1}}$ and $X_2^{h^{-1}}$ are Y-conjugate unless either

(i) the row of the table giving the embeddings involves a parameter r, s, t or q, or

(ii) Y has a factor D_m with $m = 4, 6$ or 8, and the embeddings involve the representations $3 \otimes 1^{(p)}, 5 \otimes 1^{(q)}$ or $7 \otimes 1^{(p)}$ (in which case there are two classes of A_1s in D_m with the given representation).

In case (i), the fact that $L(G) \downarrow X_1$ and $L(G) \downarrow X_2$ have the same composition factors forces the embeddings of $X_1^{g^{-1}}$ and $X_2^{h^{-1}}$ to correspond to the same values of q, r, s or t. And in case (ii), only one of the classes of A_1s in D_m gives rise to a factor $(p-2) \otimes c^{(p)}$. $\square$

Lemma 6.14 *Theorem 6.4 holds when $G = E_7$, and also when $p \geq 13$.*

Proof. Suppose that $p \geq 13$ or $G = E_7$. We know that X_1 and X_2 lie in subsystem subgroups, with embeddings as described in 6.12, and also that they of course have the same highest T_0-weight. Hence, from the table in 6.12, either the hypothesis of 6.13 holds, or $(G, p) = (E_7, 11), (E_8, 13)$ or $(E_8, 17)$, and the embedding $X_1 < Y$ is one of the following:

Y	embedding of X_1	p	highest T_0-weight
$A_1 A_5$	$1^{(p)}, 5$	11	22
$A_1 D_6$	$1^{(p)}, 6/4$	11	22
$A_1 D_6$	$1^{(p)}, 10/0$	17	34
$A_1 A_1 D_6$	$1, 1^{(p)}, 10/0$	17	34
$A_1 E_7$	$X_1 < A_1^{(p)} A_1 F_4$	17	34
$A_1 A_1 D_6$	$1^{(p)}, 1^{(p)}, 10/0$	17	44
D_7	$10/2^{(p)}$	17	44
$A_1 D_6$	$1^{(p)}, 8/2$	13	26
$A_1 A_1 D_6$	$1, 1^{(p)}, 8/2$	13	26
$A_1 D_6$	$1, 8/2^{(p)}$	13	34
$A_3 D_5$	$1 \otimes 1^{(p)}, 8/0$	13	34
$A_1 A_1 D_6$	$1^{(p)}, 1^{(p)}, 8/2$	13	34
$A_1 A_1 D_6$	$1, 1, 8/2^{(p)}$	13	34
$A_1 A_1 D_6$	$1^{(r)}, 1^{(p)}, 8/2 \, (r > p)$	13	$2r$
$A_1 A_1 D_6$	$1^{(r)}, 1, 8/2^{(p)} \, (r > p)$	13	$2r$

Suppose $p = 17$. If the highest T_0-weight is 44, then for each of the two possible mebeddings we have $X_1 < B_1^{(p)} B_5 < D_7$. Hence 6.13 applies. Otherwise, the highest weight is 34 and the three listed embeddings give different composition factors for $L(G) \downarrow X_1$: if $Y = A_1 D_6$ then $L(G) \downarrow X_1$ has a trivial composition factor; if $Y = A_1 A_1 D_6$ then $L(G) \downarrow X_1$ has just one factor $2^{(p)}$, and no trivials; and if $Y = A_1 E_7$ then $L(G) \downarrow X_1$ has two factors $2^{(p)}$ and no trivials. Hence again 6.13 gives the conclusion.

Now let $p = 13$. If the highest T_0-weight is 34 and $Y = A_1 A_1 D_6$, then $X_1 < B_1 B_1^{(p)} B_4 < A_3 D_5$. Hence either $X_1 < A_1 D_6$ or $X_1 < A_3 D_5$; only the first possibility gives trivial composition factors in $L(G) \downarrow X_1$, so 6.13 applies. For the other highest T_0-weights, we check that the different possibilities for Y lead to different composition factors for $L(G) \downarrow X_1$, so that 6.13 again applies.

Finally, if $p = 11$ then $G = E_7$ and we again check that the two possibilities for Y give different composition factors for $L(G) \downarrow X_1$. $\square$

Lemma 6.15 *Theorem* 6.4 *holds when* $G = E_8$ *and* $p = 11$.

Proof. From the table in 6.12, we see that either the hypothesis of 6.13 holds, or the embedding $X_1 < Y$ and highest T_0-weight is in the following table.

Y	embedding of X_1	highest T_0-weight
$A_1 A_5$	$1^{(p)}, 5$	22
$A_1 D_6$	$1^{(p)}, 6/4$	22
$A_1 A_1 A_5$	$1, 1^{(p)}, 5$	22
$A_1 A_2 A_5$	$1^{(p)}, 2, 5$	22
$A_1 A_1 D_5$	$1, 1^{(p)}, 8/0$	22
$A_1 A_1 D_6$	$1, 1^{(p)}, 8/2$	22
$A_1 A_1 D_6$	$1, 1^{(p)}, 6/4$	22
$A_1 D_6$	$1^{(p)}, 10/0$	26
$A_1 A_1 D_6$	$1, 1^{(p)}, 10/0$	26
$A_3 D_5$	$3, 6/2^{(p)}$	28
$A_1 A_1 D_6$	$1^{(p)}, 1^{(p)}, 6/4$	28
$D_4 D_4$	$3 \otimes 1^{(p)}, 6/0$	28
$A_1 A_1 D_6$	$1^{(p)}, 1^{(p)}, 10/0$	32
D_7	$10/2^{(p)}$	32
$A_1 D_6$	$1, 5 \otimes 1^{(p)}$	34
$A_1 A_1 D_6$	$1, 1^{(p)}, 8/2^{(p)}$	34
$A_1 A_1 D_6$	$1, 1, 5 \otimes 1^{(p)}$	34
$A_1 E_7$	$X_1 < A_1 G_2 C_3$	34
$A_1 D_6$	$1^{(p)}, 5 \otimes 1^{(p)}$	38
$A_1 A_1 D_6$	$1, 1^{(p)}, 5 \otimes 1^{(p)}$	38
$A_1 A_2 A_5$	$1^{(p)}, 2^{(p)}, 5$	44
$A_1 A_1 D_6$	$1^{(p)}, 1^{(p)}, 5 \otimes 1^{(p)}$	44
$A_1 A_1 D_6$	$1^{(p)}, 1^{(p)}, 8/2^{(p)}$	44
$A_1 A_1 A_5$	$1^{(r)}, 1^{(p)}, 5$	$2r \, (r > p)$
$A_1 A_1 D_6$	$1^{(r)}, 1^{(p)}, 10/0$	$2r \, (r > p)$
$A_1 A_1 D_6$	$1^{(r)}, 1^{(p)}, 6/4$	$2r \, (r > p)$
$A_1 A_1 D_6$	$1^{(r)}, 1^{(p)}, 5 \otimes 1^{(q)}$	$2r \, (r > q)$
$A_1 A_1 D_6$	$1^{(r)}, 1, 5 \otimes 1^{(p)}$	$2r \, (r > p)$
$A_1 A_2 A_5$	$1^{(p)}, 2^{(r)}, 5$	$4r \, (r > p)$
$A_1 A_1 D_6$	$1^{(r)}, 1^{(p)}, 5 \otimes 1^{(r)}$	$4r \, (r > p)$

If the highest T_0-weight is 28, then in all cases $X_1 < B_1^{(p)} B_2 B_3 < D_4 D_4$ and so 6.13 applies. And if the highest weight is 32 then $X_1 < B_1^{(p)} B_5 < D_7$ in both cases, and again 6.13 applies.

Suppose now that the highest weight is $4r \, (r > p)$. Let $X_1 < A_1 A_2 A_5$ (with embedding $1^{(p)}, 2^{(r)}, 5$), so $X_1 < A_1 E_7$, and let $X_1^{(0)}$ be the projection of X_1 in the

E_7 factor. Then

$$L(E_7) \downarrow X_1^{(0)} = (2^{(r)} \otimes 8)^2/(2^{(r)} \otimes 4)^2/4^{(r)}/(2^{(r)})^3/10/8/6/4/2.$$

If $X_0 < A_1 D_6 < E_7$ with embedding $1^{(r)}, 5 \otimes 1^{(r)}$ in $A_1 D_6$, we find that $L(E_7) \downarrow X_0$ has the same composition factors as $L(E_7) \downarrow X_1^{(0)}$. Hence by 6.3 and induction, $X_1^{(0)}$ and X_0 are E_7-conjugate, and so $X_1 < A_1 A_1 D_6$ with embedding $1^{(r)}, 1^{(p)}, 5 \otimes 1^{(r)}$. Thus 6.13 applies in this case. A similar argument applies to the first two entries with highest weight 44.

When the highest T_0-weight is 22, the method of the previous paragraph shows that the cases $Y = A_1 A_2 A_5$ (embedding $1^{(p)}, 2, 5$) and $Y = A_1 A_1 D_6$ (embedding $1, 1^{(p)}, 6/4$) give conjugate subgroups X_1. Similarly, when the highest weight is 34, $Y = A_1 A_1 D_6$ (embedding $1, 1, 5 \otimes 1^{(p)}$) and $Y = A_1 G_2 C_3$ give conjugate subgroups X_1. Apart from these, all the remaining possibilities for Y give different composition factors for $L(G) \downarrow X_1$. Hence 6.13 applies. $\square$

Lemmas 6.5, 6.8, 6.9, 6.14 and 6.15 complete the proof of Theorem 6.4.

7 Further results: Theorems 3 and 8

In this section we establish the two theorems remaining to be proved, Theorems 3 and 8.

Proof of Theorem 3

Let G be a simple exceptional group in characteristic p, and suppose that X is a simple closed connected subgroup of G of rank at least 2. Assume that $p = 0$ or $p > N(X, G)$ and also that p is a good prime for G. Then (X, G) is as in Tables 8.1-8.5.

The restriction $L(G) \downarrow X$ is given in the tables; the composition factors occurring are those of $L(X)$, together with those of the Weyl modules $W_X(\lambda)$ for λ as listed in the tables. If $L(X)$ and the Weyl modules $W_X(\lambda)$ occurring all have no trivial X-composition factors, then we see by inspection of the tables that the number of trivial composition factors in $L(G) \downarrow X$ is equal to the dimension of $C_G(X)$, and hence

$$C_{L(G)}(X) = L(C_G(X)).$$

When $L(X)$ or some $W_X(\lambda)$ has a trivial composition factor, we see from the tables that the only possibilities for (X, Y, p) with p a good prime (and Y a minimal subsystem group containing X) are $(A_6, A_6, 7)$, $(B_3, A_6, 7)$ and $(A_4, A_4, 5)$. In each case, there is just one extra trivial composition factor in $L(G) \downarrow X$, arising from the 1-dimensional space $Z(L(Y))$; thus if t is the number of trivial composition factors in $L(G) \downarrow X$, then $\dim C_G(X) = t - 1$.

We now show that in these three cases we still have $C_{L(G)}(X) = L(C_G(X))$. Setting $V = V_Y(\lambda_1)$, we can regard $L(Y)$ as a Y-submodule of codimension 1 in $V \otimes V^*$ (see 1.1). Now X fixes a unique 1-space of bilinear forms on V, hence fixes a unique 1-space in $V \otimes V^*$. Hence if $X = Y$ then $L(Y) \downarrow X$ is indecomposable, with a unique fixed 1-space; and if $X = B_3, Y = A_6$ then $L(Y) \downarrow X = L(X) \oplus W$, where W is indecomposable with a unique fixed 1-space. Since $L(G)$ is self-dual, it follows that $L(G)$ has a nontrivial X-submodule M such that $(L(G)/M) \downarrow X$ is indecomposable with a trivial composition factor at the top. It follows that $C_{L(G)}(X)$ has dimension $t - 1$, and hence $C_{L(G)}(X) = L(C_G(X))$, as required.

Proof of Theorem 8

We begin by defining the labelled diagram of an arbitrary closed connected simple subgroup X of the exceptional simple group G, following [Se2, §2]. Let T_0 be a maximal torus of X, and let $\Sigma(X)$ be the corresponding root system of X. Choose a fundamental system $\Pi(X)$ in $\Sigma(X)$, and denote by $\Sigma(X)^+$ the set of positive roots relative to $\Pi(X)$. If $\alpha \in \Sigma(X)^+$, and U_α is the corresponding root subgroup of X, then $\langle U_{\pm\alpha} \rangle$ is an image of SL_2, and we write $h_\alpha(c)$ for the image of the matrix diag(c, c^{-1}) $(c \in K^*)$. For $c \in K^*$, define

$$T(c) = \prod h_\alpha(c),$$

where the product is taken over all $\alpha \in \Sigma(X)^+$. Then

$$T_X = \langle T(c) : c \in K^* \rangle$$

is a 1-dimensional torus of X.

Now let T be a maximal torus of G containing T_0, and let $\Sigma = \Sigma(G)$ be the root system of G relative to T. For $\beta \in \Sigma$, let e_β be a weight vector for T corresponding to β. Then there exist integers l_β such that for all $\beta \in \Sigma, c \in K^*$,

$$T(c)e_\beta = c^{l_\beta}e_\beta.$$

As in Section 6, we choose a fundamental system $\Pi = \Pi(G)$ such that $l_\alpha \geq 0$ for all $\alpha \in \Pi$, and define the *labelled diagram* of X to be the Dynkin diagram of G, with each node $\alpha \in \Pi$ labelled by the non-negative integer l_α. By Proposition 6.2, the labelled diagram of X is uniquely determined (up to graph automorphisms of G).

Notice that, in the case where X contains a regular subgroup A_1 (which occurs provided p is sufficiently large), the labelled diagram of X is just the labelled diagram of this A_1, since T_X is a maximal torus of this A_1. Also, if Y is a subsystem group containing X, then any Aut Y-conjugate of X has the same labelled diagram as X (up to graph automorphisms of G).

Suppose that X_1 and X_2 are closed connected simple subgroups of G of the same type and of rank at least 2. Assume that $p = 0$ or $p > N(X_1, G)$ and that X_1 and X_2 have the same labelled diagram. Choose minimal subsystem groups Y_i containing X_i $(i = 1, 2)$, so that the embeddings $X_i \leq Y_i$ are as in Table 8.1-8.5.

Consider first the case where $G = E_8$. If Y_1 is conjugate to Y_2, say $Y_1 = Y_2^g$, and the embeddings $X_1 \leq Y_1$, $X_2 \leq Y_2$ correspond to the same line of Table 8.1, then by inspection of Table 8.1, X_1 and X_2^g are Aut Y_1-conjugate, as in the conclusion of Theorem 8. So assume that this is not the case. In the table on the following page, for each possibility $X_1 < Y_1$ given by Table 8.1, we list the corresponding highest weight of T_{X_1} on $L(G)/L(X_1)$ with its multiplicity (sometimes we also give the second highest weight).

We see from the table that different possibilities for $X_1 \leq Y_1$ give different T_{X_1}-weights, except in the case where $X_1 \cong A_2$ and $Y_1 = A_2A_5$ or D_4D_4. Since the labelled diagrams of X_1 and X_2 are the same, so are the T_{X_1}- and T_{X_2}-weights. Therefore it must be the case that $X_1 \cong X_2 \cong A_2$ with $Y_1 = A_2A_5$, $Y_2 = D_4D_4$. This is the exception in the conclusion of Theorem 8 (that it is a genuine exception to the general statement of Theorem 8 can be seen from Table 8.1, since application of a graph automorphism of Y_1 to X_1 puts it in D_4D_4).

This completes the proof of Theorem 8 for $G = E_8$. The other exceptional groups G are treated in exactly the same way; the relevant tables for $G = E_7$ and E_6 are given after the E_8 table (and we leave the calculations for F_4, G_2 to the reader).

Highest T_{X_1}-weights for $G = E_8$

X_1	Y_1	highest T_{X_1}-weight(s) on $L(G)/L(X_1)$
A_7	A_7	15
	A_7'	16
A_4	A_4	6^{10}
	$A_4 A_4$	10^4
	$A_4 A_4^q$	$8q$
B_4	D_5	10^8
	D_8	18^2
C_4	E_6	16
	A_7	15^2
D_4	D_4	6^{24}
	$D_4 D_4$	12^3
	$D_4 D_4^q$	$10q$
A_3	A_3	4^{10}
	$A_3 A_3$	7^4
	$A_3 A_3^q$	$6q, (4q+3)^2$
	$A_3 A_3'$	8
	$A_3 A_3'^q$	$6q, 4q+4$
	D_8	10^4
B_3	D_4	6^{25}
	$D_4 D_4$	12^3
	$D_4 D_4^q$	$10q$
C_3	A_5	9^2
	D_7	14
G_2	D_4	6^{26}
	$D_4 D_4$	12^3
	$D_4 D_4^q$	$10q$
	D_7	18
	E_6	16
B_2	A_3	4^{11}
	$A_3 A_3$	7^4
	$A_3 A_3^q$	$6q, (4q+3)^2$
	$A_3 A_3'$	8
	$A_3 A_3'^q$	$6q, 4q+4$
	$A_3 A_4$	9^2
	$A_3 A_4^q$	$8q, (6q+3)^2$
	$A_3^q A_4$	$6q, (4q+4)^2$

B_2, ctd.	$A_4 A_4$	10^4
	$A_4 A_4^q$	$8q, (6q+4)^2$
	D_5	10
	D_8	$7q+3$
	$A_3 D_5$	$6q', 4q'+4q$
	$A_3 D_5^q$	$10q$
	$A_3^q D_5$	$6q, 4q+6$
	D_7	14
	E_8	18^2
A_2	A_2	2^{54}
	$A_2 A_2$	4^{13}
	$A_2 A_2^q$	$4q, (2q+2)^{12}$
	$A_2 A_2 A_2$	6^2
	$A_2 A_2 A_2^q$	$4q, (2q+4)^2$
	$A_2^q A_2^q A_2$	$(4q+2)^2$
	$A_2 A_2^q A_2^{q'}$	$4q', (2q'+2q+2)^2$
	$A_2 A_2 A_2 A_2$	6^8
	$A_2 A_2 A_2 A_2^q$	$4q, (2q+4)^6$
	$A_2^q A_2^q A_2^q A_2$	$(6q)^2, (4q+2)^6$
	$A_2 A_2 A_2^q A_2^q$	$(4q+2)^4$
	$A_2 A_2 A_2^q A_2^{q'}$	$4q', (2q'+2q+2)^4$
	$A_2 A_2^q A_2^q A_2^{q'}$	$4q', (2q'+4q)^2$
	$A_2 A_2^q A_2^{q'} A_2^{q''}$	$4q'', (2q''+2q'+2q)^2$
	A_5	8
	$A_2 A_5$	8^3
	$A_2 A_5^q$	$8q$
	$A_2^q A_5$	$4q, (2q+6)^2$
	$D_4 D_4$	8^3
	$D_4 D_4^q$	$(6q)^2, (4q+4)^3$
	E_6	10^2
	$A_2 E_6$	10^4
	$A_2 E_6^q$	$(10q)^2$
	$A_2^q E_6$	$4q, (2q+8)^2$
	E_7	16

Highest T_{X_1}-weights for $G = E_7$

X_1	Y_1	highest T_{X_1}-weight(s) on $L(G)/L(X_1)$
A_5	A_5	8^6
	A_5'	9^2
D_4	D_4	6^{12}
	A_7	12^3
A_3	A_3	4^6
	$A_3 A_3$	8
	$A_3 A_3^q$	$6q$
B_3	D_4	6^{13}
	A_6	12^3
C_3	A_5	8^7
	A_5'	9^2
G_2	D_4	6^{14}
	A_6	12^3
	E_6	16
B_2	A_3	4^7
	$A_3 A_3$	8
	$A_3 A_3^q$	$6q$
	D_5	10
A_2	A_2	2^{30}
	$A_2 A_2$	4^9
	$A_2 A_2^q$	$4q, (2q+2)^8$
	$A_2 A_2 A_2$	6^2
	$A_2 A_2 A_2^q$	$4q, (2q+4)^2$
	$A_2^q A_2^q A_2$	$(4q+2)^2$
	$A_2 A_2^q A_2^{q'}$	$4q', (2q'+2q+2)^2$
	A_5	$8, 6^8$
	A_5'	$8, 6^8$
	$A_2 A_5$	8^3
	$A_2 A_5^q$	$8q$
	$A_2^q A_5$	$4q, (2q+6)^2$
	E_6	10^2
	E_7	16

Highest T_{X_1}-weights for $G = E_6$

X_1	Y_1	highest T_{X_1}-weight(s) on $L(G)/L(X_1)$
A_3	A_3	4^4
	A_5	8
G_2	D_4	6^8
	E_6	16
B_2	A_3	4^5
	A_4	8
	$D_5(\text{red.})$	$6q$
	$D_5(\text{irred.})$	10
A_2	A_2	2^{18}
	$A_2 A_2$	4^7
	$A_2 A_2^q$	$4q, (2q+2)^6$
	$A_2 A_2 A_2$	6^2
	$A_2 A_2 A_2^q$	$4q, (2q+4)^2$
	$A_2^q A_2^q A_2$	$(4q+2)^2$
	$A_2 A_2^q A_2^{q'}$	$4q', (2q'+2q+2)^2$
	A_5	8
	E_6	10^2

8 Tables of simple subgroups of rank at least 2 in exceptional groups

In this final section, we present the tables which give the conjugacy classes of simple closed connected subgroups X of an exceptional algebraic group G in characteristic p, where rank $(X) \geq 2$ and $p = 0$ or $p > N(X, G)$. The tables also give the subsystem groups Y containing X with Y minimal, the centralizers $C_G(X)^0$, and the composition factors of the restriction $L(G) \downarrow X$. The last two tables 8.6 and 8.7 give the composition factors of $V_{56} \downarrow X$ and $V_{27} \downarrow X$, where $V_{56} = V_{E_7}(\lambda_7)$, $V_{27} = V_{E_6}(\lambda_1)$, and X is a simple subgroup of E_7 or E_6 as in Tables 8.2, 8.3. (Note that V_{27} is not self-dual, so the entries in Table 8.7 are determined only up to a graph automorphism of X.)

Notation in the tables

1. Each row of each of Tables 8.1-8.5 corresponds to precisely one (Aut G)-conjugacy class of subgroups X, except where otherwise indicated.

2. In the column giving $L(G) \downarrow X$, each listed weight λ stands for the composition factors of the Weyl module $W_X(\lambda)$; and we write the symbol † next to λ when $W_X(\lambda)$ is reducible for some allowed value of p - the composition factors of $W_X(\lambda)$ in such cases are given after Table 8.7.

3. We sometimes write "/duals" in the $L(G) \downarrow X$ column to indicate that for any listed weight λ such that $W_X(\lambda)$ is not self-dual, the composition factors of the dual $W_X(\lambda)^*$ should also be included in $L(G) \downarrow X$.

4. In the second column of each table we give the minimal subsystem groups Y containing X; we write a superscript q, q' or q'' by a factor of Y to indicate that the projection of X to this factor involves a Frobenius morphism σ_q, $\sigma_{q'}$ or $\sigma_{q''}$ (where σ_q is a q-power morphism, and $1 < q < q' < q''$).

5. In the column giving $C_G(X)^0$, we write a bar over a factor to indicate that this factor contains fundamental subgroups SL_2 of G.

6. For groups X of rank 4 or more, we write weights λ as combinations of fundamental weights λ_i, while for groups of rank 3 or 2, we write instead abc or ab to represent the weight $a\lambda_1 + b\lambda_2 + c\lambda_2$ or $a\lambda_1 + b\lambda_2$.

Table 8.1: $G = E_8$

X	minl. subsystem groups containing X	$C_G(X)^0$	$L(G) \downarrow X$
A_8	A_8	1	$\lambda_1 + \lambda_8\,^\dagger/\lambda_3/\lambda_6$
D_8	D_8	1	$\lambda_2\,^\dagger/\lambda_7$
A_7	A_7	T_1	$\lambda_1 + \lambda_7\,^\dagger/\lambda_1/\lambda_2/\lambda_3/\lambda_5/\lambda_6/\lambda_7/0$
	A_7'	$\bar{A}_1$	$\lambda_1 + \lambda_7\,^\dagger/\lambda_2^2/\lambda_6^2/\lambda_4/0^3$
B_7	D_8	1	$\lambda_2\,^\dagger/\lambda_1\,^\dagger/\lambda_7$
D_7	D_7	T_1	$\lambda_2\,^\dagger/\lambda_1^2/\lambda_6/\lambda_7/0$
E_7	E_7	$\bar{A}_1$	$\lambda_1\,^\dagger/\lambda_7^2/0^3$
A_6	A_6	$\bar{A}_1 T_1$	$\lambda_1 + \lambda_6\,^\dagger/\lambda_1^3/\lambda_2^2/\lambda_3/\lambda_4/\lambda_5^2/\lambda_6^3/0^4$
B_6	D_7	A_1	$\lambda_2\,^\dagger/\lambda_1^3\,^\dagger/\lambda_6^2/0^3$
D_6	D_6	$\bar{A}_1^2$	$\lambda_2\,^\dagger/\lambda_1^4/\lambda_5^2/\lambda_6^2/0^6$
E_6	E_6	$\bar{A}_2$	$\lambda_2\,^\dagger/\lambda_1^3/\lambda_6^3/0^8$
A_5	A_5	$\bar{A}_1 \bar{A}_2$	$\lambda_1 + \lambda_5\,^\dagger/\lambda_1^6/\lambda_2^3/\lambda_3^2/\lambda_4^3/\lambda_5^6/0^{11}$
B_5	D_6	$\bar{B}_2$	$\lambda_2\,^\dagger/\lambda_1^5\,^\dagger/\lambda_5^4/0^{10}$
D_5	D_5	$\bar{A}_3$	$\lambda_2\,^\dagger/\lambda_1^6/\lambda_4^4/\lambda_5^4/0^{15}$
A_4	A_4	$\bar{A}_4$	$\lambda_1 + \lambda_4\,^\dagger/\lambda_1^{10}/\lambda_2^5/\lambda_3^5/\lambda_4^{10}/0^{24}$
	$A_4 A_4$	1	$(\lambda_1 + \lambda_4)^2\,^\dagger/\lambda_1/\lambda_2/\lambda_1 + \lambda_2\,^\dagger/$ $\lambda_1 + \lambda_3\,^\dagger/$duals
	$A_4 A_4^q$ (2 classes of subgroups X)	1	$\lambda_1 + \lambda_4\,^\dagger/(\lambda_1 + \lambda_4)^{(q)}\,^\dagger/\lambda_1 \otimes \lambda_2^{(q)}/$ $\lambda_2 \otimes \lambda_4^{(q)}/$duals, and $\lambda_1 + \lambda_4\,^\dagger/(\lambda_1 + \lambda_4)^{(q)}\,^\dagger/\lambda_1 \otimes \lambda_3^{(q)}/$ $\lambda_2 \otimes \lambda_1^{(q)}/$duals
B_4 $(p \neq 2)$	D_5	$\bar{B}_3$	$\lambda_2/\lambda_1^7/\lambda_4^8/0^{21}$
	A_8, D_8	1	$\lambda_2/2\lambda_1\,^\dagger/\lambda_3^2$
	D_8	1	$\lambda_2/\lambda_3/\lambda_1 + \lambda_4\,^\dagger$
C_4 $(p \neq 2)$	E_6, A_7'	$\bar{A}_2$	$2\lambda_1/\lambda_2^6/\lambda_4\,^\dagger/0^8$
	A_7	A_1	$2\lambda_1/\lambda_1^4/\lambda_2^3/\lambda_3^2\,^\dagger/0^3$
D_4 $(p \neq 2)$	D_4	$\bar{D}_4$	$\lambda_2/\lambda_1^8/\lambda_3^8/\lambda_4^8/0^{28}$
	$D_4 D_4, A_7'$	$\bar{A}_1$	$\lambda_2^5/2\lambda_1/2\lambda_3/2\lambda_4/0^3$
	$D_4 D_4, A_7$	T_1	$\lambda_2^3/\lambda_1^2/2\lambda_1/(\lambda_3 + \lambda_4)^2/0$
	$D_4 D_4$	1	$\lambda_2^2/\lambda_1/\lambda_3/\lambda_4/\lambda_1 + \lambda_3/\lambda_3 + \lambda_4/\lambda_1 + \lambda_4$
	$D_4 D_4^q$ (3 classes)	1	see Prop. 2.1
F_4	E_6	$\bar{G}_2$	$\lambda_1\,^\dagger/\lambda_4^7\,^\dagger/0^{14}$

A_3	A_3	$\bar{D}_5$	$101/100^{16}/001^{16}/010^{10}//000^{45}$
$(p \neq 2)$	$A_3 A_3$	$A_1 T_1$	$101^4/100^6/010^6/200/110^{2\dagger}/000^4/\text{duals}$
	$A_3 A_3^q$	T_2	$101/101^{(q)}/100^2/010^2/(100^{(q)})^2/$
			$(010^{(q)})^2/100 \otimes 100^{(q)}/100 \otimes 010^{(q)}/$
			$100 \otimes 001^{(q)}/010 \otimes 100^{(q)}/000^2/\text{duals}$
	$A_3 A_3', A_5$	$\bar{A}_1 \bar{A}_2$	$101^7/010^{12}/200^2/002^2/020^{\dagger}/000^{11}$
	$A_3 A_3'^q$	$\bar{A}_1^2$	$101/101^{(q)}/010^4/(010^{(q)})^4/(100 \otimes 100^{(q)})^2/$
			$(100 \otimes 001^{(q)})^2/010 \otimes 010^{(q)}/000^6/\text{duals}$
	D_8	1	$101^2/210/012/111^{2\dagger}$
B_3	D_4	$\bar{B}_4$	$010/100^9/001^{16}/000^{36}$
$(p \neq 2)$	$D_4 D_4, A_6, A_7'$	$\bar{A}_1 T_1$	$010^5/100^6/200^{\dagger}/002^2/000^4$
	$D_4 D_4, A_7$	T_1	$010^3/100^3/001^4/002/101^{2\dagger}/000$
	$D_4 D_4^q, D_7$	T_1	$010/100^2/(100^{(q)})^2/010^{(q)}/100 \otimes 100^{(q)}/$
			$(001 \otimes 001^{(q)})^2/000$
	$D_4 D_4^q$	1	$010/100/100^{(q)}/010^{(q)}/001/001^{(q)}/$
			$100 \otimes 001^{(q)}/001 \otimes 100^{(q)}/001 \otimes 001^{(q)}$
C_3	A_5	$\bar{A}_1 \bar{G}_2$	$200/100^{14}/010^7/001^2/000^{17}$
$(p \neq 2,3)$	D_7	T_1	$200/010^2/101/110^{2\dagger}/000$
G_2	D_4	$\bar{F}_4$	$01/10^{26}/00^{52}$
$(p \neq 2,3,5.7)$	$D_4 D_4, A_6$	$\bar{A}_1 A_1$	$01^5/10^{13}/20^3/00^6$
	$D_4 D_4^q, D_7 \,(\text{red.})$	A_1	$01/10^5/(10^{(q)})^5/01^{(q)}/(10 \otimes 10^{(q)})^3/00^3$
	$D_7 \,(\text{irred.})$	T_1	$01^3/30/11^2/00$
	E_6	$\bar{A}_2$	$01/20^6/11/00^8$
B_2	A_3	$\bar{B}_5$	$02/10^{11}/01^{32}/00^{55}$
$(p \neq 2,3,5)$	$A_3 A_3$	B_2	$02^6/10^{10}/01^{16}/11^4/00^{10}$
	$A_3 A_3^q$	$A_1 A_1$	$02/10^3/(10^{(q)})^3/01^6/(01^{(q)})^6/02^{(q)}/$
			$(10 \otimes 01^{(q)})^2/(01 \otimes 10^{(q)})^2/(01 \otimes 01^{(q)})^4/00^6$
	$A_3 A_3', A_4$	$\bar{A}_4$	$02^{11}/10^{20}/20/00^{24}$
	$A_3 A_3'^q, D_5\,(\text{red.})$	$\bar{A}_3$	$02/10^6/(10^{(q)})^6/02^{(q)}/10 \otimes 10^{(q)}/$
			$(01 \otimes 01^{(q)})^8/00^{15}$
	$A_3 A_4$	A_1	$02^6/10^3/01^6/20^3/03^2/11^4/00^3$
	$A_3 A_4^q$	T_1	$02/10/01^2/(10^{(q)})^2/(02^{(q)})^3/20^{(q)}/$
			$(10 \otimes 10^{(q)})^2/(01 \otimes 10^{(q)})^2/(01 \otimes 02^{(q)})^2/00$
	$A_3^q A_4, A_3 D_5$	T_1	$02^3/10^{(q)}/(01^{(q)})^2/10^2/02^{(q)}/20/$
	$(X \leq B_2 B_2 B_2^q)$		$(10 \otimes 10^{(q)})^2/(10 \otimes 01^{(q)})^2/(02 \otimes 01^{(q)})^2/00$
	$A_4 A_4, A_3 D_5$	1	$02^6/10^4/20^2/12^4$
	$(\text{irred. in } D_5)$		
	$A_4 A_4^q, D_8$	1	$02/20/20^{(q)}/02^{(q)}/(10 \otimes 02^{(q)})^2/(02 \otimes 10^{(q)})^2$
	$D_5 \,(\text{irred.})$	$\bar{A}_3$	$02^7/11^8/12/00^{15}$
	$D_8 \,(\text{irred.})$	1	$02/02^{(q)}/10 \otimes 02^{(q)}/02 \otimes 10^{(q)}/$
			$01 \otimes 11^{(q)}/11 \otimes 01^{(q)}$

	$A_3 D_5$ $(X \leq B_2 B_2^q B_2^{q'})$	1	$02/10/10^{(q)}/10^{(q')}/02^{(q)}/02^{(q')}/$ $10 \otimes 10^{(q)}/10 \otimes 10^{(q')}/10^{(q)} \otimes 10^{(q')}/$ $(01 \otimes 01^{(q)} \otimes 01^{(q')})^2$
	$A_3 D_5^q$ (irr. in D_5)	1	$02/10/12^{(q)}/(02^{(q)})^2/10 \otimes 02^{(q)}/$ $(01 \otimes 11^{(q)})^2$
	$A_3^q D_5$ (irr. in D_5)	1	$02^2/10^{(q)}/02^{(q)}/12/02 \otimes 10^{(q)}/$ $(11 \otimes 01^{(q)})^2$
	D_7	T_1	$02/20^2/22^{\dagger}/13^{2\,\dagger}/00$
	E_8	1	$02/06/32$
A_2 $(p \neq 2,3,5)$	A_2	$\bar{E}_6$	$11/10^{27}/01^{27}/00^{78}$
	$A_2 A_2$	$\bar{A}_2 \bar{G}_2$	$11^8/10^{21}/01^{21}/20^3/02^3/00^{22}$
	$A_2 A_2^q$	$\bar{A}_2 \bar{A}_2$	$11/11^{(q)}/10^9/(10^{(q)})^9/(10 \otimes 01^{(q)})^3/$ $(10 \otimes 10^{(q)})^3/00^{16}/$duals
	$A_2 A_2 A_2, D_4$	$\bar{D}_4$	$11^{25}/30/03/00^{28}$
	$A_2 A_2 A_2$	$\bar{G}_2$	$11^9/10^8/20^7/21/00^{14}/$duals
	$A_2 A_2 A_2^q$ (2 classes)	$\bar{G}_2$	$11^8/11^{(q)}/(10 \otimes 10^{(q)})^7/20 \otimes 01^{(q)}/$ $00^{14}/$duals, and $11^8/11^{(q)}/(10 \otimes 01^{(q)})^7/20 \otimes 10^{(q)}/$ $00^{14}/$duals
	$A_2^q A_2^q A_2$ (2 classes)	$\bar{G}_2$	$11/(11^{(q)})^8/(10 \otimes 10^{(q)})^7/10 \otimes 02^{(q)}/$ $00^{14}/$duals, and $11/(11^{(q)})^8/(10 \otimes 01^{(q)})^7/10 \otimes 20^{(q)}/$ $00^{14}/$duals
	$A_2 A_2 A_2^q$	$\bar{A}_2$	$11^2/10^3/20^3//10^{(q)}/11^{(q)}/(10 \otimes 10^{(q)})^3/$ $(10 \otimes 01^{(q)})^3/11 \otimes 10^{(q)}/00^8/$duals
	$A_2^q A_2^q A_2$	$\bar{A}_2$	$11/(10^{(q)})^3/(11^{(q)})^2/(20^{(q)})^3/10/(10 \otimes 10^{(q)})^3/$ $(10 \otimes 01^{(q)})^3/10 \otimes 11^{(q)}/00^8/$duals
	$A_2 A_2^q A_2^{q'}$ (4 classes)	$\bar{A}_2$	see Prop. 2.2
	$A_2 A_2 A_2 A_2, A_2 D_4$	T_2	$11^8/10^6/20^3/21^3/30/00^2/$duals
	$A_2 A_2 A_2 A_2^q, A_2^q D_4$	T_2	$11^7/30/(10^{(q)})^3/11^{(q)}/(11 \otimes 10^{(q)})^3/$ $00^2/$duals
	$A_2^q A_2^q A_2^q A_2, A_2 D_4^q$	T_2	$11/(11^{(q)})^7/30^{(q)}/10^3/(10 \otimes 11^{(q)})^3/$ $00^2/$duals
	$A_2 A_2 A_2 A_2^q$	1	$11^3/11^{(q)}/10^2/20/21/10^{(q)}/10 \otimes 10^{(q)}/$ $10 \otimes 01^{(q)}/20 \otimes 10^{(q)}/20 \otimes 01^{(q)}/$ $11 \otimes 10^{(q)}/$duals
	$A_2^q A_2^q A_2^q A_2$	1	$11/(11^{(q)})^3/10/(10^{(q)})^2/20^{(q)}/21^{(q)}/$ $10 \otimes 10^{(q)}/10 \otimes 01^{(q)}/10 \otimes 20^{(q)}/$ $01 \otimes 20^{(q)}/10 \otimes 11^{(q)}/$duals

$A_2 A_2 A_2^q A_2^q$ (2 classes)	1	$11^2/(11^{(q)})^2/10/10^{(q)}/10 \otimes 10^{(q)}/$ $10 \otimes 01^{(q)}/20 \otimes 10^{(q)}/10 \otimes 02^{(q)}/$ $10 \otimes 11^{(q)}/11 \otimes 10^{(q)}/\text{duals, and}$ $11^2/(11^{(q)})^2/10/10^{(q)}/10 \otimes 10^{(q)}/$ $10 \otimes 01^{(q)}/20 \otimes 01^{(q)}/10 \otimes 20^{(q)}/$ $10 \otimes 11^{(q)}/11 \otimes 01^{(q)}/\text{duals}$
$A_2 A_2 A_2^q A_2^{q'}$ (4 classes)	1	see Prop. 2.2
$A_2 A_2^q A_2^q A_2^{q'}$ (4 classes)	1	see Prop. 2.2
$A_2 A_2^q A_2^{q'} A_2^{q''}$ (8 classes)	1	see Prop. 2.2
A_5	$\bar{A}_1 \bar{A}_2$	$11/20^6/21^3/30^2/22/00^{11}/\text{duals}$
$A_2 A_5$	$\bar{A}_1$	$11^2/10^2/21^3/20/30^2/22/31/00^3/\text{duals}$
$A_2 A_5^q$ (2 classes)	$\bar{A}_1$	$11/11^{(q)}/22^{(q)}/(30^{(q)})^2/(10 \otimes 02^{(q)})^2/$ $10 \otimes 21^{(q)}/00^3/\text{duals, and}$ $11/11^{(q)}/22^{(q)}/(30^{(q)})^2/(10 \otimes 20^{(q)})^2/$ $10 \otimes 12^{(q)}/00^3/\text{duals}$
$A_2^q A_5$ (2 classes)	$\bar{A}_1$	$11/11^{(q)}/22/30^2/(02 \otimes 10^{(q)})^2/$ $21 \otimes 10^{(q)}/00^3/\text{duals, and}$ $11/11^{(q)}/22/30^2/(20 \otimes 10^{(q)})^2/$ $12 \otimes 10^{(q)}/00^3/\text{duals}$
$D_4 D_4, A_7$	A_1	$11^8/30^5/03^5/22^3/00^3$
$D_4 D_4, A_7', A_2 A_5$	$\bar{A}_1$	$11^8/30^5/03^5/22^3/00^3$
$D_4 D_4^q, A_8$ (2 classes)	1	$11/30/11^{(q)}/30^{(q)}/(11 \otimes 11^{(q)})^3/\text{duals}$ (same for both classes)
E_6	$\bar{A}_2$	$11/41/14/22^6/00^8$
$A_2 E_6$	1	$11^2/21/31/41/32/\text{duals}$
$A_2 E_6^q$	1	$11/11^{(q)}/41^{(q)}/10 \otimes 22^{(q)}/\text{duals}$
$A_2^q E_6$	1	$11/11^{(q)}/41/22 \otimes 10^{(q)}/\text{duals}$
E_7	$\bar{A}_1$	$11/60/06/44^{\dagger}/00^3$

Table 8.2: $G = E_7$

X	minl. subsystem groups containing X	$C_G(X)^0$	$L(G) \downarrow X$
A_7	A_7	1	$\lambda_1 + \lambda_7{}^\dagger/\lambda_4$
A_6	A_6	T_1	$\lambda_1 + \lambda_6{}^\dagger/\lambda_1/\lambda_3/\lambda_4/\lambda_6/0$
D_6	D_6	$\bar{A}_1$	$\lambda_2{}^\dagger/\lambda_5^2/0^3$
E_6	E_6	T_1	$\lambda_2{}^\dagger/\lambda_1/\lambda_6/0$
A_5	A_5	$\bar{A}_2$	$\lambda_1 + \lambda_5{}^\dagger/\lambda_2^3/\lambda_4^3/0^8$
	A_5'	$\bar{A}_1 T_1$	$\lambda_1 + \lambda_5{}^\dagger/\lambda_1^2/\lambda_2/\lambda_3^2/\lambda_4/\lambda_5^2/0^4$
B_5	D_6	$\bar{A}_1$	$\lambda_2{}^\dagger/\lambda_1{}^\dagger/\lambda_5^2/0^3$
D_5	D_5	$\bar{A}_1 T_1$	$\lambda_2{}^\dagger/\lambda_1^2/\lambda_4^2/\lambda_5^2/0^4$
A_4	A_4	$\bar{A}_2 T_1$	$\lambda_1 + \lambda_4{}^\dagger/\lambda_1^4/\lambda_2^3/\lambda_3^3/\lambda_4^4/0^9$
B_4 $(p \neq 2)$	D_5	$\bar{A}_1 A_1$	$\lambda_2/\lambda_1^3/\lambda_4^4/0^6$
C_4 $(p \neq 2)$	E_6, A_7	T_1	$2\lambda_1/\lambda_2^2/\lambda_4{}^\dagger/0$
D_4 $(p \neq 2)$	D_4	$\bar{A}_1^3$	$\lambda_2/\lambda_1^4/\lambda_3^4/\lambda_4^4/0^9$
	A_7	1	$\lambda_2/2\lambda_1/2\lambda_3/2\lambda_4$
F_4	E_6	A_1	$\lambda_1{}^\dagger/\lambda_4^3{}^\dagger/0^3$
A_3 $(p \neq 2)$	A_3	$\bar{A}_3 \bar{A}_1$	$101/100^8/001^8/010^6/000^{18}$
	$A_3 A_3, A_5$	$\bar{A}_2$	$101^7/020{}^\dagger/000^8$
	$A_3 A_3, A_5'$	$\bar{A}_1 T_1$	$101^3/200^2/002^2/010^4/020{}^\dagger/000^4$
	$A_3 A_3^q$	$\bar{A}_1$	$101/101^{(q)}/(100 \otimes 001^{(q)})^2/$
	(2 classes)		$010 \otimes 010^{(q)}/000^3/\text{duals, and}$
			$101/101^{(q)}/(100 \otimes 100^{(q)})^2/$
			$010 \otimes 010^{(q)}/000^3/\text{duals}$
B_3 $(p \neq 2)$	D_4	$\bar{A}_1 \bar{B}_2$	$010/100^5/001^8/000^{13}$
	A_6, A_7	T_1	$010/100^2/200{}^\dagger/002^2/000$
C_3 $(p \neq 2)$	A_5	$\bar{G}_2$	$200/010^7{}^\dagger/000^{14}$
	A_5'	$\bar{A}_1 A_1$	$200/100^6/010^3{}^\dagger/001^2/000^6$
G_2 $(p \neq 2, 3, 5, 7)$	D_4	$\bar{C}_3$	$01/10^{14}/00^{21}$
	A_6	A_1	$01/10^5/20^3/00^3$
	E_6	T_1	$01/11/20^2/00$

B_2	A_3	$\bar{A}_1\bar{B}_3$	$02/01^{16}/10^7/00^{24}$
$(p \neq 2,3)$	A_3A_3, A_4	$\bar{A}_2T_1$	$02^7/10^8/20/00^9$
	$A_3A_3^q, D_5$ (red.)	$\bar{A}_1T_1$	$02/10^2/(10^{(q)})^2/02^{(q)}/10 \otimes 10^{(q)}/$
			$(01 \otimes 01^{(q)})^4/00^4$
	D_5 (irred.)	$\bar{A}_1T_1$	$02^3/11^4 \dagger/12/00^4$
A_2	A_2	$\bar{A}_5$	$11/10^{15}/01^{15}/00^{35}$
$(p \neq 2,3,5)$	A_2A_2	$\bar{G}_2T_1$	$11^8/10^7/01^7/20/02/00^{15}$
	A_2A_2	$\bar{A}_2A_1$	$11^4/10^9/01^9/20^3/02^3/00^{11}$
	$A_2A_2^q$	$\bar{A}_2T_1$	$11/11^{(q)}/10^3/(10^{(q)})^3/10 \otimes 10^{(q)}/$
	(2 classes)		$(10 \otimes 01^{(q)})^3/00^9/$duals, and
			$11/11^{(q)}/10^3/(10^{(q)})^3/10 \otimes 01^{(q)}/$
			$(10 \otimes 10^{(q)})^3/00^9/$duals
	$A_2A_2A_2, D_4$	$\bar{A}_1^3$	$11^{13}/30/03/00^9$
	$A_2A_2A_2$	A_1	$11^5/10^4/20^3/21/00^3/$duals
	$A_2A_2A_2^q$	A_1	$11^4/11^{(q)}/(10 \otimes 01^{(q)})^3/20 \otimes 10^{(q)}/$
	(2 classes)		$00^3/$duals, and
			$11^4/11^{(q)}/(10 \otimes 10^{(q)})^3/20 \otimes 01^{(q)}/$
			$00^3/$duals
	$A_2^qA_2^qA_2$	A_1	$11/(11^{(q)})^4/(10 \otimes 01^{(q)})^3/10 \otimes 20^{(q)}/$
	(2 classes)		$00^3/$duals, and
			$11/(11^{(q)})^4/(10 \otimes 10^{(q)})^3/10 \otimes 02^{(q)}/$
			$00^3/$duals
	$A_2A_2A_2^q$	T_1	$11^2/11^{(q)}/10/10^{(q)}/20/10 \otimes 10^{(q)}/$
			$10 \otimes 01^{(q)}/11 \otimes 10^{(q)}/00/$duals
	$A_2^qA_2^qA_2$	T_1	$11/(11^{(q)})^2/10^{(q)}/10/20^{(q)}/10 \otimes 10^{(q)}/$
			$10 \otimes 01^{(q)}/10 \otimes 11^{(q)}/00/$duals
	$A_2A_2^qA_2^{q'}$	T_1	see Props. 2.1,2.3
	(4 classes)		
	A_5	$\bar{A}_2$	$11/21^3/12^3/22/00^8$
	A_5'	$\bar{A}_1T_1$	$11/20^2/21/30^2/22/00^4/$duals
	A_2A_5	1	$11^2/20/21/31/22/$duals
	$A_2A_5^q$	1	$11/11^{(q)}/22^{(q)}/10 \otimes 21^{(q)}/$duals, and
	(2 classes)		$11/11^{(q)}/22^{(q)}/10 \otimes 12^{(q)}/$duals
	$A_2^qA_5$	1	$11/11^{(q)}/22/21 \otimes 10^{(q)}/$duals, and
	(2 classes)		$11/11^{(q)}/22/21 \otimes 01^{(q)}/$duals
	A_7, A_2A_5	1	$11^4/22^3/30/03$
	E_6	T_1	$11/41/14/22^2/00$
	E_7	1	$11/44$

Table 8.3: $G = E_6$

X	minl. subsystem groups containing X	$C_G(X)^0$	$L(G) \downarrow X$
A_5	A_5	$\bar{A}_1$	$\lambda_1 + \lambda_5$ †$/\lambda_3^2/0^3$
D_5	D_5	T_1	λ_2 †$/\lambda_4/\lambda_5/0$
A_4	A_4	$\bar{A}_1 T_1$	$\lambda_1 + \lambda_4$ †$/\lambda_1/\lambda_2^2/\lambda_3^2/\lambda_4/0^4$
B_4 $(p \neq 2)$	D_5	T_1	$\lambda_2/\lambda_1/\lambda_4^2/0$
C_4 $(p \neq 2)$	E_6	1	$2\lambda_1/\lambda_4$ †
D_4 $(p \neq 2)$	D_4	T_2	$\lambda_2/\lambda_1^2/\lambda_3^2/\lambda_4^2/0^2$
F_4	E_6	1	λ_1 †$/\lambda_4$ †
A_3 $(p \neq 2)$	A_3	$\bar{A}_1 \bar{A}_1 T_1$	$101/100^4/001^4/010^4//000^7$
	A_5	$\bar{A}_1$	$101/200^2/002^2/020$ †$/000^3$
B_3 $(p \neq 2)$	D_4	$A_1 T_1$	$010/100^3/001^4/000^4$
C_3 $(p \neq 2)$	A_5	$\bar{A}_1$	$200/100^2/010$ †$/001^2/000^3$
G_2 $(p \neq 2,3)$	D_4	A_2	$01/10^8/00^8$
	E_6	1	$01/11$ †
B_2 $(p \neq 2,3)$	A_3	$\bar{B}_2 T_1$	$02/01^8/10^5/00^{11}$
	A_4	$\bar{A}_1 T_1$	$02^5/10^2/20/00^4$
	$D_5 \, (X \subseteq B_2 B_2^q)$	T_1	$02/02^{(q)}/10 \otimes 10^{(q)}/(01 \otimes 01^{(q)})^2/00$
	$D_5 \, (X \text{ irred.})$	T_1	$02/12/11^2$ †$/00$
A_2 $(p \neq 2,3)$	A_2	$\bar{A}_2 \bar{A}_2$	$11/10^9/01^9/00^{16}$
	$A_2 A_2$	$\bar{G}_2$	$11^8/00^{14}$
	$A_2 A_2$	$\bar{A}_2$	$11^2/10^3/01^3/20^3/02^3/00^8$
	$A_2 A_2^q$ (2 classes)	$\bar{A}_2$	$11/11^{(q)}/(10 \otimes 10^{(q)})^3/00^8/\text{duals, and}$ $11/11^{(q)}/(10 \otimes 01^{(q)})^3/00^8/\text{duals}$
	$A_2 A_2 A_2, D_4$	T_2	$11^7/30/03/00^2$
	$A_2 A_2 A_2$	1	$11^3/10^2/20/21/\text{duals}$
	$A_2 A_2 A_2^q$ (3 classes)	1	$11^2/11^{(q)}/10 \otimes 01^{(q)}/20 \otimes 10^{(q)}/\text{duals, and}$ $11^2/11^{(q)}/10 \otimes 10^{(q)}/20 \otimes 01^{(q)}/\text{duals, and}$ $11^2/11^{(q)}/10^{(q)}/11 \otimes 10^{(q)}/\text{duals}$
	$A_2^q A_2^q A_2$ (3 classes)	1	$11/(11^{(q)})^2/10 \otimes 01^{(q)}/10 \otimes 20^{(q)}/\text{duals, and}$ $11/(11^{(q)})^2/10 \otimes 10^{(q)}/10 \otimes 02^{(q)}/\text{duals, and}$ $11/(11^{(q)})^2/10/10 \otimes 11^{(q)}/\text{duals}$

$A_2 A_2^q A_2^{q'}$ (4 classes)	1	see Prop. 2.1
A_5	$\bar{A}_1$	$11/22/30^2/03^2/00^3$
E_6	1	$11/41/14$

Table 8.4: $G = F_4$

X	minl. subsystem groups containing X	$C_G(X)^0$	$L(G) \downarrow X$
B_4	B_4	1	$\lambda_2{}^\dagger/\lambda_4$
$C_4\,(p=2)$	C_4	1	$2\lambda_1{}^\dagger/V_X(\lambda_4)$
D_4	D_4 (long)	1	$\lambda_2{}^\dagger/\lambda_1/\lambda_3/\lambda_4$
	D_4 (short, $p=2$)	1	$\lambda_2{}^\dagger/V_X(2\lambda_1)/V_X(2\lambda_3)/V_X(2\lambda_4)$
A_3	A_3 (long)	A_1	$101{}^\dagger/100^2/010^3/001^2/000^3$
	A_3 (short, $p=2$)	$\bar{A}_1$	$101{}^\dagger/010^2/V_X(200)^2/V_X(020)/V_X(002)^2/000^3$
$B_3\,(p \neq 2)$	B_3	T_1	$010/100^2/001^2/000$
$C_3\,(p \neq 2)$	C_3	$\bar{A}_1$	$200/001^2/000^3$
$G_2\,(p \neq 2)$	B_3	A_1	$01/10^5/00^3$
	$F_4\,(p=7)$	1	$01/V_X(11)$
$B_2\,(p \neq 2)$	B_2	$\bar{A}_1\bar{A}_1$	$02/10^4/01^4/00^6$
$A_2\,(p \neq 2,3)$	A_2 (long)	A_2	$11/10^6/01^6/00^8$
	A_2 (short)	$\bar{A}_2$	$11/20^3/02^3/00^8$
	$A_2 A_2, D_4$	1	$11^4/30/03$
	$A_2 A_2$	1	$11^2/10/01/21/12$
	$A_2 A_2^q$	1	$11/11^{(q)}/10 \otimes 02^{(q)}/01 \otimes 20^{(q)}$ and
	(2 classes)		$11/11^{(q)}/10 \otimes 20^{(q)}/01 \otimes 02^{(q)}$
	$A_2^q A_2$	1	$11/11^{(q)}/20 \otimes 01^{(q)}/02 \otimes 10^{(q)}$ and
	(2 classes)		$11/11^{(q)}/20 \otimes 10^{(q)}/02 \otimes 01^{(q)}$

Table 8.5: $G = G_2$

X	minl. subsystem groups containing X	$C_G(X)^0$	$L(G) \downarrow X$
A_2	A_2 (long)	1	$11{}^\dagger/10/01$
	A_2 (short, $p=3$)	1	$11{}^\dagger/V_X(30)/V_X(03)$

Table 8.6: Composition factors of $V_{56} \downarrow X$ for $X < E_7$

X	minl. subsystem group containing X	$V_{56} \downarrow X$
A_7	A_7	λ_2/λ_6
A_6	A_6	$\lambda_1/\lambda_2/\lambda_5/\lambda_6$
D_6	D_6	λ_1^2/λ_5
E_6	E_6	$\lambda_1/\lambda_6/0^2$
A_5	A_5	$\lambda_1^3/\lambda_3/\lambda_5^3$
	A_5'	$\lambda_1^2/\lambda_2/\lambda_4/\lambda_5^2/0^2$
B_5	D_6	$\lambda_1^{2\,\dagger}/\lambda_5/0^2$
D_5	D_5	$\lambda_1^2/\lambda_4/\lambda_5/0^4$
A_4	A_4	$\lambda_1^3/\lambda_2/\lambda_3/\lambda_4^3/0^6$
$B_4\,(p \neq 2)$	D_5	$\lambda_1^2/\lambda_4^2/0^6$
$C_4\,(p \neq 2)$	E_6	$\lambda_2^2/0^2$
$D_4\,(p \neq 2)$	D_4	$\lambda_1^2/\lambda_3^2/\lambda_4^2/0^8$
	A_7	λ_2^2
F_4	E_6	$\lambda_4^{2\,\dagger}/0^4$
$A_3\,(p \neq 2)$	A_3	$100^4/001^4/010^2/000^{12}$
	$A_3 A_3\,(2\text{ classes})$	$200/002/010^6$ and $010^4/101^2/000^2$
	$A_3 A_3^q\,(2\text{ classes})$	$010^2/(010^{(q)})^2/100 \otimes 100^{(q)}/\text{duals, and}$
		$010^2/(010^{(q)})^2/100 \otimes 001^{(q)}/\text{duals}$
$B_3\,(p \neq 2)$	D_4	$100^2/001^4/000^{10}$
	A_6	$100^2/010^2$
$C_3\,(p \neq 2)$	A_5	$100^7/001$
	A_5'	$100^4/010^{2\,\dagger}/000^4$
$G_2\,(p \neq 2,3.5,7)$	D_4	$10^6/00^{14}$
	A_6	$10^4/01^2$
	E_6	$20^2/00^2$
$B_2\,(p \neq 2,3)$	A_3	$01^8/10^2/00^{14}$
	A_4	$10^6/02^2/00^6$
	$A_3 A_3^q$	$10^2/(10^{(q)})^2/(01 \otimes 01^{(q)})^2/00^4$
	$D_5\,(\text{irred.})$	$02^2/11^{2\,\dagger}$
$A_2\,(p \neq 2,3.5)$	A_2	$10^6/01^6/00^{20}$
	$A_2 A_2\,(2\text{ classes})$	$10^7/01^7/20/02/00^2$ and $10^6/01^6/11^2/00^4$
	$A_2 A_2^q\,(2\text{ classes})$	$10^3/(10^{(q)})^3/10 \otimes 10^{(q)}/00^2/\text{duals, and}$
		$10^3/(10^{(q)})^3/10 \otimes 01^{(q)}/00^2/\text{duals}$

$A_2 A_2 A_2$ (2 classes)	$11^6/00^8$ and $10^2/01^2/20^2/02^2/11^2/00^4$
$A_2 A_2 A_2^q$ (3 classes)	$11^2/(10 \otimes 01^{(q)})^2/00^4/$duals, and
	$11^2/(10 \otimes 10^{(q)})^2/00^4/$duals, and
	$10/20/10 \otimes 10^{(q)}/10 \otimes 01^{(q)}/00^2/$duals
$A_2^q A_2^q A_2$ (3 classes)	$(11^{(q)})^2/(10 \otimes 01^{(q)})^2/00^4/$duals, and
	$(11^{(q)})^2/(10 \otimes 10^{(q)})^2/00^4/$duals, and
	$10^{(q)}/20^{(q)}/10 \otimes 10^{(q)}/10 \otimes 01^{(q)}/00^2/$duals
$A_2 A_2^q A_2^{q'}$	see Prop. 2.3
A_5	$20^3/02^3/30/03$
A_5'	$20^2/02^2/21/12/00^2$
$A_2 A_5$ (2 classes)	$11^2/30^2/03^2$ and $10/01/21/12/30/03$
$A_2 A_5^q$ (2 classes)	$30^{(q)}/10 \otimes 20^{(q)}/$duals, and $30^{(q)}/10 \otimes 02^{(q)}/$duals
$A_2^q A_5$ (2 classes)	$30/20 \otimes 10^{(q)}/$duals, and $30/20 \otimes 01^{(q)}/$duals
E_6	$22^2/00^2$
E_7	$60/06$

Table 8.7: Composition factors of $V_{27} \downarrow X$ for $X < E_6$
(determined up to graph aut. of X)

X	minl. subsystem group containing X	$V_{27} \downarrow X$
A_5	A_5	λ_1^2/λ_4
D_5	D_5	$\lambda_1/\lambda_4/0$
A_4	A_4	$\lambda_1^2/\lambda_3/\lambda_4/0^2$
$B_4\,(p \neq 2)$	D_5	$\lambda_1/\lambda_4/0^2$
$C_4\,(p \neq 2)$	E_6	λ_2
$D_4\,(p \neq 2)$	D_4	$\lambda_1/\lambda_3/\lambda_4/0^3$
F_4	E_6	$\lambda_4{}^\dagger/0$
$A_3\,(p \neq 2)$	A_3	$100^2/001^2/010/000^5$
	A_5	$010^2/101$
$B_3\,(p \neq 2)$	D_4	$100/001^2/000^4$
$C_3\,(p \neq 2)$	A_5	$100^2/010\,{}^\dagger/000$
$G_2\,(p \neq 2,3)$	D_4	$10^3/00^6$
	E_6	20
$B_2\,(p \neq 2,3)$	A_3	$01^4/10/00^6$
	A_4	$10^3/02/00^2$
	$D_5\,(X \leq B_2 B_2^q)$	$10/10^{(q)}/01 \otimes 01^{(q)}/00$
	$D_5\,(X \text{ irred.})$	$02/11^\dagger/00$
$A_2\,(p \neq 2,3)$	A_2	$10^3/01^3/00^9$
	$A_2 A_2$ (2 classes)	$10^3/01^3/11/00$, and $10^7/02$
	$A_2 A_2^q$ (2 classes)	$10^3/(01^{(q)})^3/01 \otimes 10^{(q)}$, and $10^3/(10^{(q)})^3/01 \otimes 01^{(q)}$
	$A_2 A_2 A_2$ (2 classes)	$11^3/00^3$, and $10/01/20/02/11/00$
	$A_2 A_2 A_2^q$ (3 classes)	$11/10 \otimes 01^{(q)}/01 \otimes 10^{(q)}/00$, and $11/10 \otimes 10^{(q)}/01 \otimes 01^{(q)}/00$, and $20/01/01 \otimes 10^{(q)}/01 \otimes 01^{(q)}$
	$A_2^q A_2^q A_2$ (3 classes)	$11^q/10 \otimes 01^{(q)}/01 \otimes 10^{(q)}/00$, and $11^q/10 \otimes 10^{(q)}/01 \otimes 01^{(q)}/00$, and $20^q/01^q/10 \otimes 01^{(q)}/01 \otimes 01^{(q)}$
	$A_2 A_2^q A_2^{q'}$	see Prop. 2.3
	A_5	$20^2/12$
	E_6	22

Reducible Weyl modules in Tables 8.1-8.7 (marked with †)

X	p	λ	comp. factors of $W_X(\lambda)$
A_l	$p\|l+1$	$\lambda_1 + \lambda_l$	$V_X(\lambda_1 + \lambda_l)/0$
B_l	2	λ_1	$V_X(\lambda_1)/0$
	2	λ_2	$V_X(\lambda_2)/V_X(\lambda_1)/0^{(l,2)}$
D_l	2	λ_2	$V_X(\lambda_2)/0^{(l,2)}$
E_7	2	λ_1	$V_X(\lambda_1)/0$
E_6	3	λ_2	$V_X(\lambda_2)/0$
A_4	3	$\lambda_1 + \lambda_2$	$V_X(\lambda_1 + \lambda_2)/V_X(\lambda_3)$
	2	$\lambda_1 + \lambda_3$	$V_X(\lambda_1 + \lambda_3)/V_X(\lambda_4)$
B_4	3	$2\lambda_1$	$V_X(2\lambda_1)/0$
	3	$\lambda_1 + \lambda_4$	$V_X(\lambda_1 + \lambda_4)/V_X(\lambda_4)$
C_4	2	$2\lambda_1$	$V_X(2\lambda_1)/V_X(\lambda_2)/0^2$
	3	λ_4	$V_X(\lambda_4)/0$
	3	λ_3	$V_X(\lambda_3)/V_X(\lambda_1)$
F_4	2	λ_1	$V_X(\lambda_1)/V_X(\lambda_4)$
	3	λ_4	$V_X(\lambda_4)/0$
A_3	3	110	$V_X(110)/V_X(001)$
	3	020	$V_X(020)/000$
	3	111	$V_X(111)/V_X(200)/V_X(002)$
	5	111	$V_X(111)/V_X(010)$
B_3	7	200	$V_X(200)/000$
	7	101	$V_X(101)/V_X(001)$
C_3	7	110	$V_X(110)/V_X(001)$
	3	010	$V_X(010)/000$
B_2	7	13	$V_X(13)/V_X(03)$
	7	22	$V_X(22)/V_X(02)$
	5	11	$V_X(11)/V_X(01)$
A_2	7	44	$V_X(44)/V_X(11)$
G_2	7	11	$V_X(11)/V_X(20)$

References

[And] H.H. Andersen, "Filtrations of cohomology modules for Chevalley groups", *Ann. Sci. Ec. Norm. Sup.* **16** (1983), 495-528.

[AJL] H.H Andersen, J. Jorgensen and P. Landrock, "The projective indecomposable modules of $SL(2, p^n)$", *Proc. London Math. Soc.* **46** (1983), 38-52.

[ABS] H. Azad, M. Barry and G.M. Seitz, "On the structure of parabolic subgroups", *Comm. in Alg.* **18** (1990), 551-562.

[BT] A. Borel and J. Tits, "Éléments unipotents et sousgroupes paraboliques de groupes réductifs", *Invent. Math.* **12** (1971), 95-104.

[BW] N. Burgoyne and C. Williamson, "Some computations involving simple Lie algebras", *Proc. 2nd Symp. Symbolic and Algebraic Manipulation* (ed. S.P. Petrick), N.Y. Assoc. Computing Machinery (1971), 162-171.

[Ca] R.W. Carter, "Conjugacy classes in the Weyl group", *Compositio Math.* **25** (1972), 1-59.

[CLSS] A.M. Cohen, M.W. Liebeck, J. Saxl and G.M. Seitz, "The local maximal subgroups of exceptional groups of Lie type, finite and algebraic", *Proc. London Math. Soc.* **64** (1992), 21-48.

[DL] D.I. Deriziotis and M.W. Liebeck, "Centralizers of semisimple elements in finite twisted groups of Lie type", *J. London Math. Soc.* **31** (1985), 48-54.

[Dy] E.B. Dynkin, "Semisimple subalgebras of semisimple Lie algebras", *Amer. Math. Soc. Translations* **6** (1957), 111-244.

[GL] D. Gorenstein and R. Lyons, "The local structure of finite groups of characteristic 2 type", *Memoirs Amer. Math. Soc.*, No. 276 (1983).

[Hu] J.E. Humphreys, *Introduction to Lie algebras and representation theory*, Springer-Verlag, New York, 1972.

[Ja] J.C. Jantzen, *Representations of algebraic groups*, Academic Press, 1987.

[LT] R. Lawther and D.M. Testerman, "A_1 subgroups in exceptional algebraic groups", to appear.

[LS1] M.W. Liebeck and G.M. Seitz, "Maximal subgroups of exceptional groups of Lie type, finite and algebraic", *Geom. Dedicata* **36** (1990), 353-387.

[LS2] M.W. Liebeck and G.M. Seitz, "Subgroups generated by root elements in groups of Lie type", *Annals of Math.* **139** (1994), 293-362.

[Se1] G.M. Seitz, "The maximal subgroups of classical algebraic groups", *Memoirs Amer. Math. Soc.*, No. 365 (1987).

[Se2] G.M. Seitz, "Maximal subgroups of exceptional algebraic groups", *Memoirs Amer. Math. Soc.*, No. 441 (1991).

[Sp] T.A. Springer, *Linear algebraic groups*, Progress in Math. Vol. 9, Birkhauser (Boston, Basel, Stuttgart), 1981.

[St] R. Steinberg, "Lectures on Chevalley groups", Yale University Lecture Notes, 1968.

[Te1] D.M. Testerman, "Irreducible subgroups of exceptional algebraic groups", *Memoirs Amer. Math. Soc.*, No. 390 (1988).

[Te2] D.M. Testerman, "A construction of certain maximal subgroups of the algebraic groups E_6 and F_4", *J. Algebra* **122** (1989), 299-322.

Imperial College, London SW7 2BZ, UK

University of Oregon, Eugene, Oregon 97403, USA

Editorial Information

To be published in the *Memoirs*, a paper must be correct, new, nontrivial, and significant. Further, it must be well written and of interest to a substantial number of mathematicians. Piecemeal results, such as an inconclusive step toward an unproved major theorem or a minor variation on a known result, are in general not acceptable for publication. *Transactions* Editors shall solicit and encourage publication of worthy papers. Papers appearing in *Memoirs* are generally longer than those appearing in *Transactions* with which it shares an editorial committee.

As of January 31, 1996, the backlog for this journal was approximately 5 volumes. This estimate is the result of dividing the number of manuscripts for this journal in the Providence office that have not yet gone to the printer on the above date by the average number of monographs per volume over the previous twelve months, reduced by the number of issues published in four months (the time necessary for preparing an issue for the printer). (There are 6 volumes per year, each containing at least 4 numbers.)

A Copyright Transfer Agreement is required before a paper will be published in this journal. By submitting a paper to this journal, authors certify that the manuscript has not been submitted to nor is it under consideration for publication by another journal, conference proceedings, or similar publication.

Information for Authors and Editors

Memoirs are printed by photo-offset from camera copy fully prepared by the author. This means that the finished book will look exactly like the copy submitted.

The paper must contain a *descriptive title* and an *abstract* that summarizes the article in language suitable for workers in the general field (algebra, analysis, etc.). The *descriptive title* should be short, but informative; useless or vague phrases such as "some remarks about" or "concerning" should be avoided. The *abstract* should be at least one complete sentence, and at most 300 words. Included with the footnotes to the paper, there should be the 1991 *Mathematics Subject Classification* representing the primary and secondary subjects of the article. This may be followed by a list of *key words and phrases* describing the subject matter of the article and taken from it. A list of the numbers may be found in the annual index of *Mathematical Reviews*, published with the December issue starting in 1990, as well as from the electronic service e-MATH [**telnet e-MATH.ams.org** (or **telnet 130.44.1.100**). Login and password are **e-math**]. For journal abbreviations used in bibliographies, see the list of serials in the latest *Mathematical Reviews* annual index. When the manuscript is submitted, authors should supply the editor with electronic addresses if available. These will be printed after the postal address at the end of each article.

Electronically prepared papers. The AMS encourages submission of electronically prepared papers in $\mathcal{A}_{\mathcal{M}}\mathcal{S}$-TEX or $\mathcal{A}_{\mathcal{M}}\mathcal{S}$-L^AT_EX. The Society has prepared author packages for each AMS publication. Author packages include instructions for preparing electronic papers, the *AMS Author Handbook*, samples, and a style file that generates the particular design specifications of that publication series for both $\mathcal{A}_{\mathcal{M}}\mathcal{S}$-TEX and $\mathcal{A}_{\mathcal{M}}\mathcal{S}$-L^AT_EX.

Authors with FTP access may retrieve an author package from the Society's Internet node **e-MATH.ams.org** (130.44.1.100). For those without FTP access, the author package can be obtained free of charge by sending e-mail to `pub@math.ams.org` (Internet) or from the Publication Division, American Mathematical Society, P.O. Box 6248, Providence, RI 02940-6248. When requesting an author package, please specify $\mathcal{AMS}$-TeX or $\mathcal{AMS}$-LaTeX, Macintosh or IBM (3.5) format, and the publication in which your paper will appear. Please be sure to include your complete mailing address.

Submission of electronic files. At the time of submission, the source file(s) should be sent to the Providence office (this includes any TeX source file, any graphics files, and the DVI or PostScript file).

Before sending the source file, be sure you have proofread your paper carefully. The files you send must be the EXACT files used to generate the proof copy that was accepted for publication. For all publications, authors are required to send a printed copy of their paper, which exactly matches the copy approved for publication, along with any graphics that will appear in the paper.

TeX files may be submitted by email, FTP, or on diskette. The DVI file(s) and PostScript files should be submitted only by FTP or on diskette unless they are encoded properly to submit through e-mail. (DVI files are binary and PostScript files tend to be very large.)

Files sent by electronic mail should be addressed to the Internet address `pub-submit@math.ams.org`. The subject line of the message should include the publication code to identify it as a Memoir. TeX source files, DVI files, and PostScript files can be transferred over the Internet by FTP to the Internet node `e-math.ams.org` (130.44.1.100).

Electronic graphics. Figures may be submitted to the AMS in an electronic format. The AMS recommends that graphics created electronically be saved in Encapsulated PostScript (EPS) format. This includes graphics originated via a graphics application as well as scanned photographs or other computer-generated images.

If the graphics package used does not support EPS output, the graphics file should be saved in one of the standard graphics formats—such as TIFF, PICT, GIF, etc.—rather than in an application-dependent format. Graphics files submitted in an application-dependent format are not likely to be used. No matter what method was used to produce the graphic, it is necessary to provide a paper copy to the AMS.

Authors using graphics packages for the creation of electronic art should also avoid the use of any lines thinner than 0.5 points in width. Many graphics packages allow the user to specify a "hairline" for a very thin line. Hairlines often look acceptable when proofed on a typical laser printer. However, when produced on a high-resolution laser imagesetter, hairlines become nearly invisible and will be lost entirely in the final printing process.

Screens should be set to values between 15% and 85%. Screens which fall outside of this range are too light or too dark to print correctly.

Any inquiries concerning a paper that has been accepted for publication should be sent directly to the Editorial Department, American Mathematical Society, P. O. Box 6248, Providence, RI 02940-6248.